STILL THE SAME HAWK

STILL THE SAME HAWK

Reflections on Nature and New York

Edited by
JOHN WALDMAN

Empire State Editions
An imprint of Fordham University Press
New York 2013

Fordham University Press also publishes its books in a variety of electronic formats. Some content that appears in print may not be available in electronic books.

Library of Congress Cataloging-in-Publication Data is available from the publisher.

Printed in the United States of America
15 14 13 5 4 3 2 1
First edition

Contents

Acknowledgments

This book would not exist save for the generosity of the late labor negotiator and philanthropist Theodore Kheel. Ted Kheel, as he was known, developed a passion for urban nature and sustainability near the end of his long and productive life. This interest prompted his funding of a new institute at the City University of New York (CUNY)—the CUNY Institute for Sustainable Cities (CISC). Some of the essays in this collection result from a conference held at Queens College in 2005 to launch the CUNY Institute, titled "Why Nature Matters to New Yorkers." I also thank Dr. Gillian Small, CUNY Vice Chancellor for Research, and Dr. William Solecki, the Director of CISC, for their critical support in the production of this volume. Finally, I owe special gratitude to one of the contributors, Dr. Frederick Buell, who provided his considerable wisdom with abundant patience whenever called upon throughout the editing process.

This volume is dedicated

to the late Theodore Kheel,

who saw the value in an

ecologically healthy and sustainable city.

Introduction

Bare Nature in the Naked City

John Waldman

Dualism is the defining quality of urban nature.

Few juxtapositions conjure as many mixed reactions from city dwellers—oft confused, occasionally distressed, sometimes remarkably uplifted—as the blatant appearance of "nature" against their urban backdrop.

1

This dualism also was a defining quality of my life; I was fortunate to be both Tom Sawyer *and* Huck Finn. I grew up well within New York City limits, in a private house on a busy street in the northeast Bronx, walking distance to the elevated subway and only a few doors from an expressway. All day our home vibrated from speeding trucks. It was sport in the neighborhood to listen for and to be the first kid to the scene of an accident on that road; as latecomers arrived they asked, "Anyone dead?" and "Who got there first?"—not always in that order, and with primacy affording a week of bragging rights.

But only a bike ride away was Eastchester Bay, an inelegant arm of Long Island Sound. The bay had already yielded a goodly portion of its acreage to the massive Pelham Bay Landfill, as its wetlands had yielded to the Soviet-in-its-charmlessness Co-Op City apartment projects. But to me, these environmental insults didn't much matter. What I discovered was a bay alive with wooden sailboats and cabin cruisers, wheeling sea gulls and terns, critter-filled tide pools, and boulders to be flipped at low tide to see what hid under them. There were endless flounder, snapper bluefish and, after I'd advanced in my learning, wily striped bass to be caught. Better yet, one of Gotham's best-kept open-space secrets, Pelham Bay Park, edged the bay, and after school on a snowy day I could track rabbits by their footprints in the park's woods and make a fire and cook hotdogs, making believe I was Mark Trail, Davy Crockett, Lewis or Clark, or some as yet unsung *über* urban explorer.

So I've seen New York from my dress shoes standing on the asphalt and from my sneakers sinking into the muck of the bay. This upbringing left me curious about how others viewed the mingling of nature and metropolis. I've explored this in various ways: through observing reactions to my anecdotes, through reading contemporary newspaper accounts, through books and archived articles. But perhaps most effectively, I have explored this subject through images.

2

Arthur Leipzig created the photograph "East River Divers" in 1948. There is no denying the sheer grace and almost liquid sense of movement as the boys leap in sequence. But they are plunging into New York's East River, what was then a heavily polluted tidal strait—in the image the water's surface has an off-putting, milky organic glow. Despite this, there is purpose, and gusto, and maybe even a sense of joy in their flight.

Imagine that this same arc of motion was occurring from a height over a farm pond somewhere instead of against the backdrop of New York's 59th Street Bridge. It would not be a scene that still resonates and is remembered some sixty years later; it would be long since forgotten, having no more aesthetic weightiness than a postcard. It is the stark juxtaposition between this wordless narrative and its unlikely context that renders the photograph sublime, that gives it a duality that stirs the soul. Such is the essence of nature in the city.

3

Urban legends are hatched here in the miasma of big city life, among them the fantastical notion that our sewers are home to fierce alligators, living there because owners of baby alligators jettisoned their pets once they grew too large to live in the bathtub. Travel to the 14th Street station of the 8th Avenue subway line and you'll see a cast-bronze gator emerging from a manhole and engulfing a pedestrian, as pictured on

Copyright Arthur Leipzig, 1948

page 4. Despite the fact that the human figures in the ensemble have as heads only expressionless moneybags bearing dollar signs where their faces should be, the victim comes across as clearly confused and the bystander as merely bemused. The comic rendition of the urban legend in Tom Otterness's *Life Underground* does not mask its subtext: that even when the city dweller is seemingly cocooned in concrete, latent fears intrude; we are never fully safe from the perils of the natural world.

4

The artist Saul Steinberg was a master at the efficiently ironical view of New York City. The spare sketch reproduced on page 5 appeared on the dust jacket and title page of the 1950s work *The Bottom of the Harbor* by *New Yorker* writer Joseph Mitchell. It was a perfect pairing. In his lean and lively prose Mitchell chronicled the lives and personalities of the characters who peopled the various water trades of the City, bringing the harbor's mysteries to life. These include factually based legends such as what occurs annually in mid-April; as I wrote in *Heartbeats in the Muck*,

"as the depths warm, bacterial activity will bloat the previous winter's bounty of murders and suicides and cause them to rise to the harbor's surface—a synchronized resurrection of the damned that captains call Floater's Week."

Of all the feral corners and spaces of the city, none evokes greater macabre fascination than New York Harbor. This has never been truer than in midcentury, when the legacy of almost two hundred years of gross pollution and other environmental atrocities had not yet begun to reverse in response to the Clean Water Act of 1972. Corpses still rise during Floater's Week, but they no longer drift along in fetid sludge and chemical stew.

Despite this now-lauded improvement, nature remains largely invisible underwater in the natural murk of the Hudson Estuary. For this reason, I argue that the wildest place within New York City limits is not Central Park's Ramble, Staten Island's swamps, or the outermost limits of Pelham Bay Park. The wildest place in all of New York City, surprisingly, is right at its geographical center—the bottom of Hell Gate.

Hell Gate is that sharp bend in the East River where the current speeds and boils as it glides over tall bedrock formations that rise from the bottom. On a full running tide it's still a frightening place, but its danger today pales against the torrent that existed before its most threatening reefs were blown up and excavated late in the nineteenth century.

Before that was done it reached 10 knots—roughly the average speed of a winning marathon runner—and its whirlpools could be heard whooshing a quarter hour before a vessel reached it. Many shipwrecks ensued. To the best of my knowledge, no one has yet to venture to the bottom of its 100-foot depths.

Steinberg's sketch incorporates a brilliant contrast, one that illuminates long-held public perceptions of the harbor. In it Manhattan's gorgeous spires are packed tightly from bank to bank, with the composition's visual weight balanced by Lady Liberty. The utility of the harbor is represented by a tugboat towing a cruise ship. Whatever occurs above the waterline is splendid, but below it, details are sparse: Two fish (one of them eyeless) represent life, and the essential mystery of the realm is symbolized by a human skeleton tumbling out of a 55-gallon drum.

5

Apex predators, those creatures at the top of the food chain, are among the most iconic symbols of nature. There was a time in the United States when these predators, be they wolf, wildcat, or bird of prey, were routinely killed (sometimes with the encouragement of bounty payments) under the misguided notion that this was bedrock-basic wildlife management. The terribly oversimplified guiding equation: fewer predators meant more of their prey available for people. But the city resident can claim no such rationale; instead, the appearance of one of these creatures in the urban landscape is an upset in the normal order of life—one that may provoke embarrassingly primitive responses.

Perhaps the low point in the reaction of New Yorkers to the invasion of their city by such a predator occurred in 1950. A photograph from the Brooklyn *Daily Eagle* newspaper shows a shark that wandered into the far end of Brooklyn's Gowanus Canal. In those days, the canal was still a legendarily polluted waterway, one so fouled with sewage and industrial wastes that its stench could be smelled blocks away; in fact, Joseph Mitchell wrote around that time that "only germs can live there." In earlier times captains docked their vessels in it because the waters would kill barnacles on the hulls, saving them the labor of scraping. If a human was found in the canal, he or she was dead, and had almost always been dead before hitting the water.

The photo shows a remarkable execution scene. Word had spread through the neighborhood of the shark's presence and hundreds of people had gathered. They can be viewed in the image's background as they

stand watching as a team of police sharpshooters fires, the bullet slicing the water as it strikes near the fish's fin. Instead of marveling at this unlikely visitor from the ocean, the populace of this quintessentially unnatural environment—a place where no one would ever consider swimming—had defended itself against what was instinctively, and irrationally, perceived as a menace.

6

The recent controversy over the fate of Fifth Avenue's now-famous red-tailed hawk, Pale Male, and his mate, Lola, is praiseworthy to New Yorkers and represents a new view of the apex predator. A few residents of a tony apartment building tired of the carcasses left by the hawks that nested on a ledge and also, perhaps, of all the unwanted attention focused on their dwelling, and so they had the nest dismantled. Red-tailed hawks are distinctly not cuddly; they swoop down on, strike, and bloody the same pigeons and squirrels in Central Park that many people care enough about to feed daily. And yet, instead of a firing squad, these beaked and taloned predators inspired candlelight vigils on their behalf, three books, a website, and a documentary film. The saga of this single pair of hawks demonstrates how the growing ecological awareness of the late twentieth century has reached the citizens of an environ seemingly far removed from the natural. Instead of feeling threatened by the predator, we now celebrate it.

7

In considering cities, naturalists and urbanists are in rival intellectual camps that have unsettled and evolving relationships with each other. The collection of creative nonfiction essays in this volume was written by some of the most probing analysts of what nature means to a great city—these are reflections by naturalists and urbanists that should not only inform the role of nature in New York City, but in urban centers across the world. Duality is rife throughout these works. One writer tells of a student of his who remarked, "The problem with nature in New York is that there isn't any." But another chapter makes the case that nature and New York are one and can't be separated, that the natural world "is still, despite everything, such a powerful and dynamic force that it does not even do it justice to call it the elephant in the room that nobody mentions. The elephant envelops the room."

The sudden manifestation of a natural phenomenon may enrich the daily similitude of city life. Betsy McCully, then a Brooklyn newcomer, was one of the many local residents who were "drawn by an invisible string" to the brief appearance of multitudes of monarch butterflies on a

Coney Island beach. Why do urbanites flock to spectacles of this kind? In "Monarchs of the Urban Mind" she sees the monarch phenomenon as part of a seamless whole that includes the insect, the goldenrod it feeds on, and New York City on its migratory path, but also as a relic of a deeper time, with greater plentitude, being discovered, with delight, by a populace also made up of migrants. In such a dynamic landscape, McCully's encounters with nature help orient her in a place where "everything has changed and nothing has changed."

To me, Tony Hiss is the poet laureate of the emerging discipline of a "sense of place." Recently, he was the lead author of a synthesis of New York's natural characteristics for the Dodge Foundation, titled the *Highlands to Ocean Report*. Coauthored with his colleague Christopher Meier, their chapter "Welcome to the H2O Region—Your Second Address!" follows an original point of view, making the case for a new regional identity for New York City—one that defines borders not in political terms but based on natural features. The dichotomy between the two leads the authors to implore city residents to view their home as having two addresses, the traditional one they are accustomed to and the natural one that is their place in the environment of the region.

Every developed lot and parcel in New York City was untamed once, and each has wavered between these states over its history as the city has evolved and re-invented itself. In "Public Place, Brooklyn," Kelly McMasters settles for a time in a flat across the street from an open plot that is in environmental limbo, a brownfield space that contains both a legacy of contaminants and the dreams of residents for a new beginning as a park to enjoy. But developers covet the parcel too, and the nearby shoreline of the legendarily polluted Gowanus Canal. McMasters tracks the fate of this small piece of ground long after she moves away, as factions battle to a standoff in which no one benefits except those who enjoy the feeling of openness it provides when they look beyond the razor-wire fence that now surrounds it.

If "Public Place, Brooklyn" portrays a never-realized Arcadian dream, then Dara Ross's "Corner Garden" recounts Paradise Lost. "Corner Garden" tells how local residents took the initiative to simply clean up and begin gardening another unused Brooklyn patch—no questions asked—staking plants and staking a claim, albeit temporarily, to a richer life. This fragment of nature and the commonalities of the food, recreation, community, and even the lingering scents it affords enlivened the spirit of a diverse immigrant neighborhood.

William Kornblum has studied the New York environment for his entire life, most pointedly seeing New York Harbor from the viewpoints of a dock laborer in his youth and, later, from that of a respected sociologist who explores the harbor via his vintage catboat. In "A Land Ethic for the City" Kornblum discusses Aldo Leopold and his influential land ethic, but he argues for a new *urban* land ethic, one in which city residents, though they don't own land on the scale of Leopold's farmers and ranchers, nonetheless recognize their tremendous stake in the lands and waters that contain them and their city. Kornblum demonstrates the power of this new way of thinking by drawing from his experiences teaching city-dwelling undergraduates about the environmental issues and possibilities for responsible stewardship that surround them.

Essayist Phillip Lopate is a staunch urbanist and thus offers a compelling counterpoint to the other contributors in this collection. Instead of asking only "Why does nature matter to New Yorkers?" he says that "maybe the question to be asked is: Why should New York and other big cities matter to naturalists?" In "Can Naturalists and Urbanists Find Happiness Together?" Lopate accepts the value of nature in the urban setting, but makes a case for letting cities be cities—for recognizing that cities possess still-underrecognized value as an efficient means to harbor large numbers of people in limited areas, thus preventing more widespread environmental harm.

With support from the late philanthropist Theodore Kheel, David Rosane spent time immersing himself in the environment, history, and lore of nature in New York City with the charge of sensitizing city dwellers to its natural wonders. And with a background in tropical biology and a bent towards philosophy, Rosane is a fresh, one-of-a-kind voice for urban nature, capable of drawing connections among biodiversity, nonlinear dynamics, and voodoo chickens. His offbeat chapter "Can You Eat in Soup? Nine Million Ways to Look at a Raccoon—and an Apple" examines many of the ways humans interact with nature in the city environment, with its hub a revealing story about the breadth of New Yorker's reactions to stumbling upon a raccoon in Central Park.

Cities offer among the starkest contrasts between the natural and unnatural. Some of these may be so off-putting as to reach almost (urban) mythological status. Robert Sullivan has been an original student of two of the darker manifestations of nature in New York: the long-sundered Hackensack Meadowlands and the place of the lowly rat in city life. In "The Dark Side; or, My Time Spent in the Nature That People Would Rather Not Think About," Sullivan uses these subjects, comically but also with great insight, to explore the contrasts between the compromised nature of a metropolis and the idealized nature of wilderness.

The New York City region faces numerous threats, including continued suburban sprawl, climate change, and sea-level rise. Anne Matthews, a Midwesterner excited by her personal revelation of the tenacity of nature in the city, examines these threats in "The Futures of New York" with a sober yet poetic point of view, asking "What are the futures of New York?" Matthews proposes four possibilities that range from "a dystopian battleground of civic chaos and decay" to a "deindustrializing city morphed into a sustainable one." None are definitive, but all are plausible, and, whatever the outcome, the New York City of the future will be viewed—with its stark nature/culture contrasts but also its remarkably small ecological footprint—as a model for an urbanizing planet.

Ecocriticism is an analysis of how texts construct human and nonhuman relations to nature. Devin Zuber maintains that there is an abiding presumption that nature is out there, beyond the city, in the great national parks of the West or upstate, but certainly not in a metropolis such as New York City. In "Imagination, Beauty, and the Urban Land Ethic: Teaching Environmental Literature in New York City," Zuber uses the writings of the other chapter authors and of key Romantic figures—one American (Walt Whitman) and one British (William Blake)—to articulate a contemporary ecological sensibility for public service and pedagogy within the confines of the modern city.

Finally, in "Nature in New York: A Brief Cultural History," Frederick Buell portrays the writers in this collection as accomplishing something new in environmental literature—as part of a vanguard movement in nature writing, one increasingly important not just for urban audiences but for future environmental representation and thought. Buell describes the important conventions of nature writing from Thoreau to the present and surveys the massive social, cultural, and environmental changes that have challenged the tradition of nature writing during the past forty years. He then goes on to propose recognition for the emergence of a new tradition of hybrid urban nature writing as one of today's most meaningful literary responses to those changes.

8

Duality again. Pale Male's eviction sparked a major contretemps. A deep grassroots passion for the fate of wild beings was apparent. City residents felt inspired by the presence of these predators making a life in such unlikely urban circumstances, and the groundswell of concern—bordering on outrage—culminated in restoration of their nest. And yet, in one of this book's essays, the author suggests that a hawk that lives on an apartment building is no more unusual than a hawk that lives on a cliff. It's still the same hawk.

Monarchs of the Urban Mind

Betsy McCully

On a cloudless September Sunday in 1984, thousands of monarch butterflies descended from the sky to nectar on seaside goldenrods, their collective weight bending the stalks down. One might expect migrating monarchs in a country field, but not in the rock rubble of a Manhattan Beach jetty. Rumors flew around my Coney Island neighborhood, and people rushed down to the impromptu monarch festival, myself among them. Monarchs clustered so densely they turned the goldenrods orange. They hovered above the flowers, each waiting its turn to unfurl its proboscis into the nectar sacs and sip. They floated above our heads and alighted in our hair and on outstretched hands as children joyfully reached out to them. For two days they remained, nectaring and resting before resuming their migration southward; then just as mysteriously as they had appeared they vanished, all but a few stragglers reminding us of what had been.

It was my first year living in New York City, and I just assumed that the monarchs stopped here every year, like the swallows of Capistrano. But in the years since, they have never returned in such numbers. It was a rare event, not because monarchs are rare, but because it was a random and ephemeral occurrence. And it was unexpected because it happened in a densely populated, heavily polluted, and radically altered urban landscape.

Coney Island was once a place of dunes, beach grasses, and bayberries, divided from Long Island by a tidal creek. It's thought that Henry Hudson stopped here before piloting his Half Moon into New York Bay. The

Dutch named the island after the rabbits that abounded here, the *konijen*, or coneys. Development of the island began in the 1820s, but not until after the Civil War was its oceanfront built up as a resort and pleasure ground. In the 1870s, real estate developer August Corbin bought up land on Coney Island's eastern tip, a swath of salt marsh known as Sedge Bank, renaming it Manhattan Beach. He built two resort hotels—the Oriental and the Manhattan Beach—along the shorefront. To protect the resorts from the sea's onslaught, men constructed rock jetties and sea walls. They made an esplanade so people could stroll from Manhattan Beach to Brighton Beach and take the sea air. They flattened the dunes to make beaches. The hotels have long given way to houses, and over the decades, hurricanes and nor'easters have broken up the jetty and esplanade. Goldenrods and other wild plants that can withstand salt spray have anchored their roots in pockets of soil.

The monarchs have imprinted themselves in my memory not as isolated objects to net and mount for study, but as parts of a living whole. Monarchs gliding in the medium of air, warming their bodies in the sun, fueling on nectar for their flight. Goldenrods harnessing energy in the solar cells of their leaves, taking up minerals and water in their stiff stems, transforming light and water into flowers that thrust their stamens and pistils toward pollinators to ensure their reproduction. And people drawn by invisible strings toward the monarch gathering, becoming a part of the scene, entering the monarchs' universe. How improbable the monarchs seem in this urban landscape! Yet they are merely repeating a migratory cycle begun thousands of years ago at the end of the last ice age, when a warming climate drew this tropical butterfly northward.

Long before New Yorkers hardened the city's shorelines and pumped sand onto smoothed and flattened beaches, monarchs nectared on goldenrods that grew in coastal dunes. They are one of a handful of butterfly species that migrate, and they travel the longest distance, two thousand miles between Canada and Mexico. For countless generations they have obeyed an inner GPS, migrating southward in fall and northward in spring. On their north-bound journey, they nectar on milkweed blossoms, mate, lay eggs on the underside of milkweed leaves, and die. Their larvae hatch, feed on the milkweed leaves, and pupate into butterflies that continue their journey northward. On their southward journey, the fourth generation returns to their wintering grounds in the Sierra Madre of Mexico, where they roost in stands of Oyamel trees. Lengthening days and warming temperatures in spring trigger their reproductive cycle, and a new generation sets out on their journey north. How each successive generation knows to follow the same migratory routes is still one of nature's mysteries.

I envy the monarch's unerring sense of direction and homing instinct. Typical of a global megacity, New York is a city of immigrants, whether first-, second-, or tenth-generation removed. The difference between our human migrations and those of monarchs and other migratory species is our linear trajectories. Most of us do not follow a cyclical pattern timed to the seasons. When I came to New York, I had moved a dozen times and lived in six cities in a restless lifetime. I was tired of my peripatetic wanderings and longed for a home ground. I wanted to put down roots here and get to know the place where I live. My chance encounter with the monarchs started me on a different kind of journey, at once intellectual and spiritual, toward a deeper, more personal knowledge of place—this urban habitat I call home. Just as I wanted to anchor myself, I sought to anchor my city in its bioregion, however severely degraded by our human imprint that bioregion may be. To understand that process of degradation, it became necessary to reverse time, to roll the film backward to see the different stages of the city's growth until I came to the land that was before the city. Then I began to mentally excavate the layers of time and space ever deeper, delving into a billion years of cataclysmic geological events. What I gained from this deep-time perspective was a sense of place that is dynamic and ever evolving.

Here, continents docked and rifted, mountains rose and eroded, glaciers advanced and retreated. It's a place where land has always met sea, however shifting the shoreline. Flash back to twenty thousand years ago: a glacier as high as the Statue of Liberty scraped Manhattan's bedrock and piled debris into moraines. Coney Island and Long Island were parts of a vast coastal plain, a landscape of tundra and spruce bogs threaded with streams and rivers that sluiced glacial outwash toward the sea, then a hundred miles distant. Flash back to two hundred million years ago: the earth rifted as eastern North America split from western Africa, marking the breakup of the supercontinent Pangaea. Molten rock erupted through fissures in the earth's crust, hardening into blocks of diabase sill that eons later would be tilted up and exposed as the 400-foot-tall cliffs known as the Palisades on the New Jersey side of the Hudson. Directly across from the Palisades rise the cliffs of Manhattan's northern tip. These schist outcrops attest to earth-shaking events of four hundred million years ago, when eastern North America collided and fused with volcanic offshore islands, pushing up colossal mountains. The eroded roots of those mountains form Manhattan's bedrock. Flash forward to the present: at the turn of the last century, when they excavated foundations for skyscrapers and tunneled for subways, engineers used the ripped-up bedrock, called rip-rap, to build and reinforce the city's jetties. Now, as sea level rises in a rapidly warming world, waves churned up by storms crash over the jetties, and the sea surges under Coney Island's boardwalk, sucking away sand that must be continually replenished for beachgoers.

In addition to my research into New York's natural history, I have also explored the city as it is, here and now—a human construct that is never fixed. I have walked as much of the city as I can to know its terrain. My historical perspective enables me to traverse both past and present. Sidewalks beneath me dissolve into shorelines, graded streets into wetlands, concrete hills into forested uplands. I've taken copious notes on the natural life I've observed wherever I could find it, and through all my wanderings I've cultivated a deepening awareness of urban nature, an openness to chance encounters, a fine-tuning of my senses to catch the nature that surrounds us in the city. Or is it that nature catches us?

Looking back through two decades of nature journals, I'm struck by what captured me: always a juxtaposition of wild and human constructs, a surprising insertion of the wild into human space—or is it that I am inserted into wild space? Or an interplay of both? A dweller in an urban habitat, I am embedded in nature by an act of mind that connects me to the life web.

One fog-bound morning I stepped out of my house and nearly into a spiderweb. An orb weaver had spun its web across the driveway, threading a silken strand from bush to car antenna to another bush, constructing a foot-wide web. Its shimmer caught my eye, each strand pearled with water droplets that refracted the light. I was brought up short, my breath stopped.

It was a chance encounter with the wild—yet the spider was not so alien to me in my constructed world. It was doing what it had evolved to do—to weave webs to catch prey. I was doing what I as a human had evolved to do—think, reflect on the spider and its web, and imagine the spider's world. By my act of consciousness, I engage in a process as natural as the spider's spinning of its web. And yet—we blunder into spiderwebs and break them. We build our cities in ways that erase nature, blindly destroying habitats as we make our own. This comes about because we have mentally and spiritually severed our connection to wildness. I believe we can reconnect to nature by opening our minds to the possibilities of wildness in our midst, even in an urban landscape, and by opening our souls to unexpected, unscripted moments of beauty.

Wildness is not necessarily rare or exotic. It may be quite commonplace—a spider or butterfly. Some of us, including myself, travel to the ends of the earth to track down the rare animal or plant, but miss the daily natural phenomena taking place by our front stoops. If we open ourselves to the possibility, we may have a surprise encounter with wildness that for a moment breaks our preoccupations, stops us in our tracks—or gets us running to catch what we can before they that wildness vanishes into thin air.

Photo by Betsy McCully

My encounters with wild animals have often happened when I was not looking for them. While I was eating breakfast one winter morning, a hawk landed on a bench in our postage-stamp backyard, so close I could see through the window his golden eye. When I pulled aside the kitchen curtain to get a better look, he flew off, fanning his barred tail. He was gone before I could pull out my bird guide and determine whether it was a Cooper's or Sharp-shinned hawk, both accipiters that visit bird feeding stations in winter—but unlikely visitors to an urban backyard. Another surprise visitor was a whimbrel ambling down the sidewalk in late summer. Shorebirds that breed in the Arctic tundra and migrate to their wintering grounds in Chile and Brazil, whimbrels normally feed on invertebrates in tidal mudflats, and here he was probing his long down-curved bill into a neighbor's lawn.

Further afield, while strolling along the Hudson esplanade in Bay Ridge one May evening, I observed horseshoe crabs lumbering onto a little mudflat that had formed next to a decayed dock beneath the Verrazano Bridge. Males slowly clambered over each other and onto females. Mating only during the new and full moons of late May and early June, horseshoes can be found performing their annual spring rites up and down the East Coast. Males patrol shallow bay waters, waiting for females to come ashore at low tide. Males may line up behind females in chains, or several at once may climb onto a female. The male clasps the larger female with his pedipalp, an appendage specially designed for the purpose, fertilizing her eggs as she lays them in holes she has prepared—thousands of eggs in multiple clutches between the low and high tide lines.

The once-abundant horseshoe, a so-called living fossil that has survived two hundred million years, is now endangered by overharvesting, its population crashing—bringing down with it the red knots, shorebirds that feed on horseshoe crab eggs to fuel themselves on their northward migratory journey from the tip of South America to Canada. Like the monarch's dependency on milkweeds and goldenrods, the red knot's dependency on horseshoe crabs reminds us of our interconnectedness—break one strand in the web and we destroy the whole structure in a cascade of extinctions.

The horseshoes, which inhabit the shallow waters of the continental shelf, remind us that the dominant element of New York City is water; in fact, the city is built on an archipelago set within a vast estuary—one of the most productive in the world. In spring, when snow melts in the Adirondacks and waters run high, the Hudson turns cocoa brown with outflowing sediments. For millennia these sediments have fed the shores of New York City, building mudflats and nourishing wetlands that have supported a teeming diversity of life. When the tide is at ebb, the Hudson flows to the sea, and at high tide the sea flows upriver. The continual mixing of fresh and salt waters in the Lower Hudson has created an estuary, a complex ecosystem that serves as a vital nursery for marine life. The extraordinary fertility of these waters has attracted humans to the region for thousands of years.

Despite our abuses—polluting the waters, obliterating the wetlands, hardening the shores—it astounds me to see how tenacious life can be. I have taken several boat trips up Newtown Creek, a heavily industrialized waterway that divides Queens from Brooklyn. Once a meandering tidal creek called *Mespacht kil* by the Dutch after the Lenape place name, in the early decades of the twentieth century it carried nearly as much freight traffic as the Mississippi. From the mid-1800s onward it was the center of the region's petrochemical industry, and although a few heavy industries still operate here, most have close their doors, leaving in their wake a toxic mess. A decades-old underwater oil spill and the overflow of raw sewage have further poisoned the waters, earning the creek its designation in 2010 as a Superfund site. Riding the boat up the creek, I feel like a tourist of urban industrial catastrophe. The boat propeller stirs up a sludgy, malodorous mass called *black mayonnaise.* In some parts of the creek, the level of dissolved oxygen has fallen to zero. It's a wonder there's any life left here—yet I see cormorants, herons, and egrets feeding. A kingfisher dives into the water. Where old bulkheads are crumbling, tiny mudflats are forming and marsh grasses taking hold. Ivy grows over a vacant brick warehouse, and sunflowers erupt like fireworks on abandoned industrial sites.

Nature is not limited to organic life. Think of wind and tide, sky and clouds, moon and stars. And light. Emily Dickinson wrote of a "certain slant of light" on snow-covered landscapes, but I see it on city buildings in winter, when everything is as stripped down as leafless trees and the air has a sharp clarity. Once, on a bright January day, the light slipped into my mind. I had crossed the blaring traffic of 42nd Street in Manhattan and entered the quiet, open space of Bryant Park, behind the New York Public Library. The space was enclosed by tall buildings, their angular towers silhouetted against the translucent sky. A row of plane trees, stirred by the wind, cast lean dancing shadows onto the paving stones. I became aware of the interplay of branching shadows and lozenges of silvery light through which I walked, myself a body casting a shadow and absorbing light. I was a walker in the city and a walker on the earth, anchored by gravity to a spinning planet orbiting the sun and cycling through the seasons.

In observing changes in the angle of light as the earth's pole tilts toward and away from the sun, I become connected to a cyclical rhythm that has been missing in my wanderings from city to city. Here, in a city of uprooted and transplanted peoples, I find it possible to reorient myself, and, like the monarchs, find my home ground. My shoes hit hard pavement, but in my mind I feel the crush of decayed leaves on a forest floor, hear the rustle of meadow grasses, smell the pungent odor of a mudflat.

I know the landscape that was here before the city stamped its grid of streets, poured its cement foundations, bulkheaded shores, filled in marshes, buried streams, flattened hills and dunes. I also know the urban landscape, where monarchs float over blaring traffic and hawks roost on high-rise ledges. Everything has changed and nothing has changed. The ocean still withdraws when the tide is at its lowest ebb, exposing mudflats to air, releasing the pungent odor of decay. Leaves still fall from trees in autumn, scattering on city sidewalks and collecting in puddles. My experience of the urban landscape is at once of past and present—deeply immersed in the present, and acutely aware of layers of history down through deep time.

On the Coney Island boardwalk, I walk through throngs of people speaking in myriad languages and accents, their voices melding with the murmuring of the waves. I cannot separate past from present, for all is a continuous process. Waves of people migrate to these shores, seeking a new home. The sea rises and expands, warmed by our human activities on this planet. Our jetties and sea walls barely withstand the sea's onslaught, and here and there appears a chink in the wall, a shifting of the esplanade, a rearrangement of rocks. And in a pocket of soil, its seed long dormant, a goldenrod takes root and flowers, inviting migrating monarchs to stop and feed before continuing their journey.

Welcome to the H2O Region—Your Second Address!

Christopher Meier and Tony Hiss

Not long ago we published a book—*H2O: Highlands to Ocean*—trying to call attention to an astonishing, seemingly unlikely, and often almost invisible fact: the natural world of the New York–New Jersey metropolitan region, site of relentless growth and development for almost four hundred years and now home to sixteen million people, is still, despite everything, such a powerful and dynamic force that it does not even do it justice to call it the elephant in the room that nobody mentions. The elephant envelops the room.

Our book both presents strong scientific evidence to back up this claim and explains how we, the inhabitants of this region, could have contrived a situation where something so vast and of such potential importance to our daily lives could disappear from view. *H2O* points to a combination of forces: we hid nature behind buildings and roads, and we trained our minds to pay more attention to the geometric lines we invented—lot lines, town lines, county lines, and state lines—than to the pre-existing and surviving ridgelines, river banks, marshlands, meadows, and forests.

Our ten-year quest to uncover, recover, and celebrate the area began by asking two more modest questions: Can the existing residents of this place find a greater sense of connection to the creatures and landscapes—the nature—that surrounds us? And, so as not to begin by assuming the facts: Is there still any nature—even a semi-bedraggled ecosystem—left to relate to?

These last two questions arose during a small, ecumenical conference held in 1997, where twelve New Yorkers and another dozen New Jerseyans parked their animosity and lingering suspicion of one another and gathered to discuss their shared heritage. We invite readers of these pages to engage in a similar exercise, by casting their gaze beyond the borders of the Big Apple itself. Allow New York City to be one element of a far larger natural area—a region spanning 5,300 square miles of land and water, an area including all or part of nine New York and twelve New Jersey counties.

This vast landscape extends 75 miles north-to-south and 70 miles east-to-west. It includes, to be sure, all of the amazing natural highlights of New York City: New York Harbor, Jamaica Bay, Central Park, and the islands poking through the Arthur Kill on the west side of Staten Island, where many harbor herons make their home. Opening one's gaze to the broader region reveals similarly astounding beauty, and in vast acreages, such as the Hackensack Meadowlands, the Glacial Lake Passaic wetlands in New Jersey, the Highlands in New Jersey and New York, and Sprain Ridge Park in Westchester County, New York.

At the gathering, the New York contingency, recognizing the region as a young, postglacial, water-sculpted and water-based place, wanted to focus on practical concerns, with statements such as, "It's the water, stupid!" The New Jerseyans were the dreamers, and asked the question, practical in its own way: How can anyone relate to a place that doesn't even have a name? One attendee, Barbara Lawrence, came to the rescue, coining the name *Highlands to Ocean*—celebrating two of the outstanding features of the region and allowing for the catchy nickname *H2O.* The meeting also led to the H2O map, and, finally, to the first comprehensive look at the health of the environment in this urban region.

The work by the two dozen people at this meeting carried over and spread to include dozens more people and groups working to bring awareness of the H2O region to all of its citizens. And now this fact can be demonstrated and proclaimed: though often hidden by the canyons of skyscrapers and rivers of concrete and obscured by the strings of power and communication lines, the region is a vast and rich natural area. A remarkably resilient landscape and waterscape survives and thrives here. As we learn more about this urban natural area, curiosity and wonder reawaken. To be clear, re-absorbing this underlying reality only begins with the readings in these pages; the next step is to explore the natural areas themselves, opening oneself up to the presence of what once was and of what still is.

Having now taken this journey ourselves, we can state with conviction that New Yorkers and New Jerseyans all have two addresses—a street address within their local community, and a place of honor within the larger H2O region. Whether at night your head rests on a pillow in a Queens bungalow or in a Manhattan high-rise 300 feet in the air, without even knowing it yet you live in an amazing natural neighborhood. Such a wonder deserves a proper introduction; allow us in the rest of this chapter to serve as tour guides to a few highlights of the H2O region. It is our hope that these words encourage and challenge you to find a pair of good walking shoes or hiking boots and head for the areas that are covered by the essays in this book, so that while there you can discover some unchronicled gems all on your own.

Parks are no doubt the first thing to enter anyone's mind at the beginning of a discussion of urban nature. Six generations ago, in the 1850s, came the first great urban parks movement, with hundreds built throughout the region. New York City has many wonderful parks, including the world's most famous one, Central Park. Now, parks may not be the wildest of areas, but they can provide important underpinnings to an ecosystem. Central Park, for example, serves as an essential annual stopover for migrating birds. The Ramble, in particular, has an international reputation as a "warbler trap"—as many as thirty species of warblers have been seen there on a single May day.

This is because New York City sits in the middle of the "great bend" of the Eastern seaboard—where the north–south-running coastline coming up from Florida becomes the east–west-running New England coast. This "corner effect" in the middle of the coast makes New York City and the larger H2O region an important hub on the Atlantic flyway, with hundreds of millions of migrating birds moving through it twice a year. Every spring and every fall, our area is an outstanding place for bird watching, which happens to be a wonderful way to begin or deepen an appreciation for the natural patterns within the urban design.

Nestled alongside Manhattan, only blocks from Central Park, is the Hudson River. It is easy to forget that water is the defining characteristic of the H2O region; our surroundings have been shaped and sculpted by the relentless and furious yet subtle and delicate carvings of ice and water flow over countless millennia, most dramatically in the form of the Hudson. The magnificent, beautiful, and world-famous Hudson River is already more than 250 miles long before it twists through the Hudson Highlands and enters the H2O region. From just above Albany south, the Hudson is "the river that flows two ways," as Native Americans

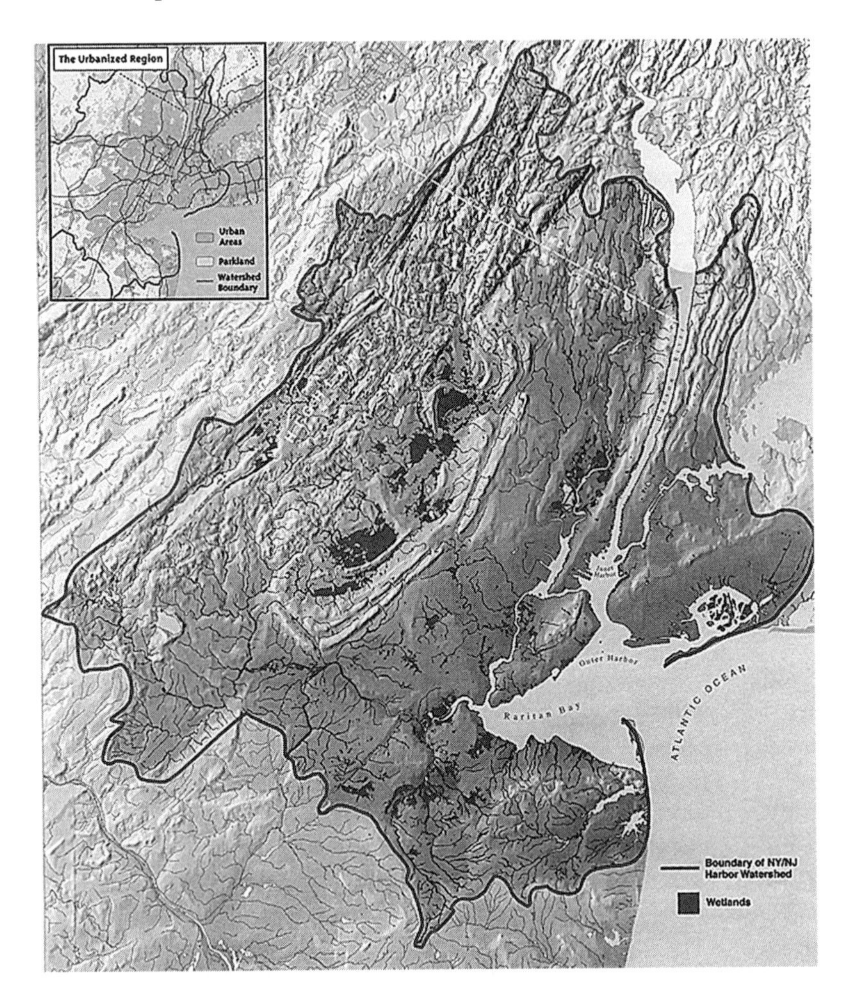

called it, a fjord-like arm of the Atlantic Ocean subject to tides that blend fresh and salt water. As such, it is an incredibly rich aquatic environment, "a kind of Times Square," its greatest biographer, Robert Boyle, has written, since it offers shelter to both marine and freshwater fish and serves as the spawning ground and nursery for important migratory fish such as striped bass and sturgeon that will spend most of their lives far at sea. Some surprises occur in Manhattan's Hudson waters: delicate seahorses nibble food from wooden pier pilings, and in late summer, juveniles of tropical fishes carried by the Gulf Stream are found, such as grey snapper, jack crevalles, and butterflyfish.

Near the mouth of the Hudson estuary lie the shallow, tranquil waters, stiff sea breezes, extensive salt marshes, and mudflats scattered with 10-foot-high islands that make up Jamaica Bay. The bay, a 20-square-mile sheltered extension of the sea separated from the Atlantic Ocean only by narrow coastal sand dunes and barrier beaches, is sometimes saluted as the Chesapeake Bay of the North. Largest of the region's urban wildernesses, Jamaica Bay is a glacially scooped-out lagoon nearly the size of Manhattan Island. Some of its long, empty, windswept vistas suggest that the bay is an almost impossibly remote setting, far from any human habitation, while other views reveal the tiny tops of downtown Manhattan's towers glimmering on the horizon 10 miles away.

Jamaica Bay is a contradictory kind of place. It is crossed by a subway line, the famous A train, and is almost constantly subjected to the overhead roars of jets, since it is bordered on the northeast by the runways of Kennedy Airport. Yet the bottom of the bay is covered by such a rich layer of mussels and other shellfish that it reminded one reporter of thick shag carpeting. The bay supports by far the region's largest colony of diamondback terrapins, a brackish-water turtle once rampantly overharvested for the fine dining it provided. The bay also shelters more than 330 species of birds and is considered one of the premier birding locations in the United States

Jamaica Bay is almost wholly within New York City limits. But looking to the larger H2O region, the first significant find beyond the city is a wilderness area about three miles west of Central Park and ten times its size. The Hackensack Meadowlands have 8,400 acres of permanently protected wetlands—currently managed by New Jersey's state-chartered Meadowlands Environmental Trust. This is our area's big-sky country, and it has endured and survived various assaults over the years. The landfills of yesteryear are now capped and sealed, and today the Meadowlands support 225 species of birds.

The Meadowlands are a bequest of Glacial Lake Hackensack. What was once a narrow, icy, postglacial, meltwater lake that extended 40 miles from near the southern end of Staten Island to the southern end of New York's Rockland County is now the low-lying saltwater and brackish wetlands we often notice only from the Pulaski Skyway or from a New Jersey Transit train on a journey to Newark Liberty International Airport. The Meadowlands are a fascinating place, and, to learn more, read a wonderful book: *The Meadowlands: Wilderness Adventures at the Edge of a City* by Robert Sullivan, a fellow contributor to this volume.

Exploring the Meadowlands on your own is becoming easier with the opening of New Jersey Transit's Secaucus Transfer train station. In earlier years, the Meadowlands were easy to get through but hard to get to. Now a train ride of seven minutes from Manhattan, or eight minutes from Newark, on the "train to the green," makes a visit to the edge of this urban wilderness a lunchtime possibility. One worthwhile next step would be an exhibit of Meadowlands murals within the train station, reminding hurrying travelers that they are not just on their way somewhere—they have already arrived at the heart of a spectacular urban H2O wilderness.

Urban wilderness areas can take some getting used to. In the Meadowlands, the sight of a waving thicket of reeds coexists with the loud traffic roars from the New Jersey Turnpike. The senses must process paradoxical information. It happens in Central Park when a warbler's song—usually audible only deep in the heart of big woodlands—can be heard, while in plain sight are the high towers of the Upper West Side skyline. Sometimes even a single sense will receive contradictory information: egrets perched at the foot of the high-tension electric pylons are a common Meadowlands sight. The key is to shift your perspective. The nature you can see and feel is not a remnant of a lost past, but a reminder of and message from the still-vital, still-enveloping "H2O-ness" of the region.

Residents of the H2O region, or of other urban areas, shouldn't be too hard on themselves for not yet knowing more about their second address. Development and sprawl does not just hide the second address of a place: it is oblivious to this continuing reality, and pretends that the second address has been uprooted forever. The result is that, without realizing it, we live "as if" lives: as if our second address were unreachable, or had ceased to exist. If you spend time exploring the coasts and wetlands of our region, it won't be long before you run into a *No Swimming* or *No Fishing* sign on a chain-link fence; much of the land, similarly, is cordoned off by *No Trespassing* signs. Harder to see are the *No Trespassing* signs posted in our own minds. The developments and mini-malls we build turn only their blank back walls onto an adjacent wetland or creek. It's one thing to degrade or distance ourselves from the natural environment. It's even worse when we pretend it's not even there.

Knowing something actually exists can be the catalyst that turns a person walking with her head down to avoid potholes into an explorer seeking out a shining wetland in every puddle. An explorer needs a map, or many maps. Paul Cohen and Robert Augustyn have put together a fascinating collection, *Manhattan in Maps: 1527–1995.* Residents of the H2O region would be well advised to get their hands on a copy of this extraordinary book, which records both the place and the mindsets that have swept over it.

In 1776, at the beginning of the Revolutionary War, the British captured Manhattan, and they held onto it until they were forced to capitulate. Since the war was being fought elsewhere, the troops on the island had time on their hands, and set about creating an incredibly intricate map of Manhattan's topography—a map they unfortunately took home to London after their surrender. Ostensibly, this "British Headquarters Map," as it was called, provided the details British commanders would need to defend the city. Once it was rediscovered, we were able to see that it displays—in exacting detail drawn on a large scale, 1 mile = 6.5 inches—the full and original second address of Manhattan and the surrounding islands. It shows a richly diverse landscape full of hills, valleys, and streams all but unrecognizable to contemporary citizens of the city. By taking the map with them, the British left Manhattan's second address adrift and undefended, because no one else had ever compiled its intricacy and elegance into a single, memorable image.

Unrestrained by this vision, New York's grand solution for this undulating landscape was devised some thirty years later, as drawn in 1811 by cartographer John Randel Jr., creator of the "Commissioners' Plan." The Commission's purpose was to set a course for the development of the island. The result was the grid system we have today, with major avenues running north-south and side streets running east-west at right angles to the avenues.

In order to superimpose this rigid geometry on the island, most of the hills that made up the postglacial landscape of Manhattan were shaved down, providing—literally—a more level playing field for developers. These hills then became fill that was dumped into the kills (*kill* being the Old Dutch word for stream). While the nature of New York could not be erased as easily as the hills, subsequent generations of Manhattanites were lulled into inaccurately assuming that their home was essentially a place of almost beachlike flatness and aridity.

Another map a modern explorer of Manhattan should be aware of is "Viele's Water Map," versions of which were first published in 1859. The cartographer, Egbert Ludovicus Viele, the engineer-in-chief of Central Park, urgently wanted to bring attention back to the waterways, both surface and underground, of New York City. Viele thought that burying these waterways for development was dangerous to the health of the city, offering analogies to the plague years of Europe. Sanitation was his cause. Instead, as Cohen and Augustyn ironically report, "over the years the map has been put to constant use by contractors, who study it to determine whether their building sites are former riverbeds that could still flood foundations." Today's H2O-ers can celebrate the Viele map for another reason: it was the first attempt to bring together the two addresses of the place, because it showed both the grid and the hills and waterways, thereby giving equal value to the two realities.

That, then, is the trick—rebalancing our expectations so that, secure in our urban and suburban houses and apartments, we can also see, welcome, participate in, and protect the extraordinary profusion of world-class natural places and their inhabitants that persist and thrive in the midst of the world's foremost metropolitan area. This may require a new way of thinking. In the early years of the environmental movement, one catchphrase was to "think globally, act locally"—although, as a practical matter, the concerns often seemed so far removed from each other that it was hard to see a direct connection between local deeds and global needs. Now, with a greater understanding, we can add a third concern to the equation—the region—to serve as a middle ground, a missing link that reflects both local actions and global consequences. Living and cooperating regionally makes it possible to think globally and act locally.

Public Place, Brooklyn

Kelly McMasters

I turned my face away from the small, squat building, watching the cloud-white jet streams streak through the blue Brooklyn sky. Leaning against a pea-green fire hydrant, painted the same color as the house behind it, I waited for the real estate agent to return from touring the other prospective renter around the basement-level space. The ad had called it a "one-bedroom garden apartment," although after weeks of scouring listings and visiting disappointing place after disappointing place, I knew a converted basement when I saw one. But I loved the frontier feeling of the block, which sported only two lumpy residential buildings sandwiched between small factories and a short warehouse, and there was something about the open lot on the other side of the street that made it a little easier for me to breathe.

Stretching out beyond a chain link and razor-wire fence, the lot was more dust than soil and there were hardly any trees, but the size of the open space—multiple acres, I figured—was foreign in the Brooklyn landscape I had come to know. I'd lived in some of the borough's softer neighborhoods, like Park Slope and Windsor Terrace, during the previous five years, and I was struggling to cross the Gowanus Canal and make it to Carroll Gardens, closer to the city and slightly edgier, trendier. A few blocks away were fancy cheese shops and cafés, and although I would have preferred to live among them, this basement apartment was at the foot of Smith Street, where the subway burst through the concrete and roared across elevated tracks. This block was not really part of the neighborhood, but over the past few weeks it had become quite clear that this block would be the closest to Carroll Gardens my bank account would allow.

I stared past the strange casket-sized blocks of graffitied cement strewn throughout the lot and looked above the curling spiral of barbed wire that topped the six-foot fence to a section of elevated subway arcing through the sky in the distance. Two F trains shuttled towards one another, brakes squealing as they braced against the turn, scuffed silver bullets on a collision course. For a moment it looked as though one train were swallowing the other whole, and then they crossed, continuing on their separate trajectories.

The street was quiet again. I looked back at the squat one-and-a-half story house in which I was trying to rent an apartment. An old sheet was pinned unevenly across the upstairs window in front, and I wondered who lived there. Then the door to the basement opened and the man who had answered the Craigslist.org ad for the apartment before I did slunk up the cement stairs and hurried back up the street. The real estate agent waved me in.

In 1895, Citizens' Gas Company set up shop on a section of land in Brooklyn where the Gowanus Canal curves like a comma. They dug three large circles into the ground, and then dropped a swimming-pool-sized concrete storage tank in each. One of ultimately two dozen manufactured-gas plants in Brooklyn, Citizens' Gas supplied nearby homes and businesses along the Gowanus with fuel for heating, cooking, and light. In 1925, the plant was taken over by Brooklyn Union Gas Company, which expanded the site in response to Brooklyn's growing population and utility demands.

Gas was produced using a process called coal carbonization. Clumps of coal were shoveled into large beehive ovens, which allowed the black rocks to heat to just under the burning point. This released vapors, which were then collected. Once the gas was skimmed off, all that was left in the beehive ovens was coke, a grainy black substance that burned hotter than coal itself, and this by-product of the gas production was also sold off. The vapors, including a flammable mixture of methane and carbon monoxide, were then combined with petroleum products to create more methane. This gas was then pumped through large pipes underground throughout the neighborhood.

By the 1960s, manufactured gas was being replaced by natural gas, a cheaper alternative, and like many other coal gasification plants, this one hugging the curved hip of the Gowanus was shut down. The property was transferred to Keyspan but was left inactive. The weeds pushed up, and the cement disks and holding tanks and other machinery sank deeper into the dark earth.

The Monday after I moved into my new apartment, I awoke to diesel fumes and the chugging of loud engines penetrating my basement bedroom at five in the morning. What I had taken on my first, Sunday-morning visit to the apartment to be a warehouse next door was actually a commercial bakery, and the delivery trucks came at the break of each dawn to fill up with sticky buns and bread and muffins. Every half hour after that the house shook as heavy trucks made their way to and from the cement factory at the end of our short block. Out back in the massive garden, which featured a lean-to with a giant hole in the middle that the landlord said had once been used for making wine, the heady fumes from the trucks mixed with the treacly scent of cinnamon rolls.

When the landlord came over to fix a loose bar on one of my windows, I asked him about the lot across the street. *Why is it empty? Are they going to build something there?* With Smith Street's recent commercial explosion I was surprised that an open space of this size would still be untouched. John, a smooth-faced, trim man of fifty who worked as a plumber, rubbed his hands together.

"It's supposed to be a park. The state's involved, because it's polluted or something, but they are going to make a beautiful park. Now that the Gowanus is getting cleaned up, this area is gonna be like the Venice of Brooklyn. And once there is a nice park there I'll knock this building down," he said, pointing to the squat green house. "Then I'll build something big." He looked at my face, saw the concern in my now-wide eyes as I imagined packing up again, and continued, "But that won't happen for a long time."

I asked what kind of pollution was in the lot.

"I don't know. Something to do with Keyspan or coal or something. They say it's bad for you, but I used to play in there when I was a kid, before they put the fence up. We'd build forts and hideouts in there, play stickball." He shrugged. "I'm still here. And this block will be the most expensive in the neighborhood, soon. You watch."

The lot was named "Public Place" in 1974. That was when talk of plans for a park seems to have started. In 1990, the New York State Department of Environmental Conservation declared the space an Inactive Hazardous Waste Site. Coal gasification plants were proving to be a tricky problem because of the chemical soup left behind; solvents, coal tar residues, arsenic and cyanide, tar-soaked bricks and timbers, and carcinogens such as benzene, toluene, ethylbenzene, and xylene remained beneath the surface, migrating through soil and bubbling up in inky black patches.

In 1995, the Environmental Protection Agency began its Brownfield Program, and Public Place was added to more than 450,000 other properties across the country targeted by the agency's efforts to sustainably reuse and redevelop land that was otherwise too polluted to build on. The idea is that reinvesting in these sites—rather than leaving them to rot because no one wants to shoulder the expense of cleaning them up—ultimately slows environmental degradation, because developers can reuse this land, rather than tearing down trees and putting pressure on or ruining other undeveloped natural spaces. The brownfield designation made Public Place the largest open space available for development in all of Brooklyn.

I taped sheets of clear plastic over the shoebox-small bedroom windows, which puffed like sails against the diesel fumes and wind. Some nights, though, when the bakery was closed and the delivery trucks silent in the parking lot down the street, I would peel off the tape from a corner and pull back the cloudy plastic to see into the sunken square in front of my house where we kept the garbage cans.

Often I would spy Louie, a small, shrunken man with a voice like a duck and a face like a raisin. He was harmless and babbled constantly, but often left the sunken square littered with beer bottles or hypodermic needles. Once, that first summer, he must have shot up in the small cavelike privacy by my front door, and as I stepped out of the house in flip-flops I narrowly missed the needle's sharp glinting stick. Another time, my parents were visiting, and as we left the house my father pointed out one of Louie's needles nearby. *A diabetic lives next door,* I lied.

Often, though, there were faces I didn't recognize. The bakery left bags of bread out every night, and the desolate block was a place to come for food and a bit of quiet. But when there were too many, or when they hopped down from the street to curl up inside the sunken square outside my bedroom window, or when I would peel away the plastic and find myself staring at a man cupping his penis and directing a yellow stream into my garbage can, I would call the cops.

When it was just Louie, I didn't call the cops. Instead, my boyfriend would walk him to the corner and ask him to stay away. On those nights,

Louie would cry as they walked and tell of the days before the fence, the days when he and the others lived in the vacant lot in peace. When the fence went up, the homeless lost their access to Public Place and to their hideaways, the small secret spaces between the cement blocks or deep inside the mouths of wide, gaping pipes. Louie said that when the fence went up, he lost his home.

Tar contains volatile chemicals that have the ability to vaporize and be inhaled. If you kick the toe of your sneaker into the sandy silt of Public Place, or if you are a little boy rolling around in the weeds during a game of stickball, or if you are a homeless person bedding down for the night, chances are you will release these chemicals into the air.

Most of the polluted structures—the gas holders, the oil tanks, the ash dump, the coal piles—are within the fenced area across the street, but there was also a five-million-cubic-foot gas holder on my side of the street. This giant keg used to tower over all of the other buildings on the block. Its footings are still buried beneath one of the warehouses on the other side of the bakery.

The exposure pathways of PAHs and BTEX, the main pollutants at Public Place, include inhalation, skin absorption of airborne benzene, direct contact with groundwater, soil vapors, or secondary intake from crops. The legal maximum contaminant level for BTEX is 5μg/L. In 2005, when I was living next door, BTEX contamination at Public Place registered between 540 and 2,600μg/L. The highest contamination level found for PAHs was 500μg/L; the maximum contaminant level for PAHs is 0.2μg/L. This means the level of PAHs was at one time 2,500 times over the level considered safe.

Much of the BTEX sludge migrated into the Gowanus Canal, sinking to the bottom in slick, oily drops that don't burst, shattering instead into hundreds of smaller drops, like the mercury in a thermometer. I imagine the bottom of the Gowanus as a bed of glittering black pearls.

I worked on my garden at the beginning of the summer until the fat, insistent mosquitoes forced me back inside. Only a few weeks after I cleared a patch of weeds and planted some small beans from a paper packet I bought for a dollar at True Value around the corner, vibrant green vines crawled up the stakes I had leaned against the cement wall nearby. Then they kept growing, climbing the wall and sending shoots out in every direction. I harvested juicy green beans as big as plantains all summer long. When some friends invited me to their summer house one weekend, I brought along a bag of my beans and steamed them, tossed them with some garlic and butter, and served them. No one could believe they came from my garden. They were the biggest beans we'd ever seen. After a while I couldn't keep up, and the beans dropped off the vines from the pull of their weight, the pods shrinking as they dried out into thin papery wings, rotting and returning to the ground.

A few months later, I walked out my door and up the front cement stairs and saw moon men moving slowly through the weeds in the open lot across the street. I stopped and stared for a while at the group of people walking around the lot in heavy white suits. These suits covered their whole bodies and heads, complete with gloves and bobble-head helmets. They lumbered around wielding silver instruments and notebooks, planting sticks into the ground like flags, or shoveling bits of dirt here and there.

I later found out that these were Columbia University engineering students. They were testing contamination levels and measuring the landscape of the lot so that they could create final projects and presentations for a project about remediation. Professors of urban design, landscape architecture, and contaminant and remediation engineering were working with a group called the Public Place Alliance, a cooperative effort of nine local organizations, in an attempt to create some possible uses that would combine both community and commerce goals. First, though, the level of pollution needed to be determined.

I didn't know who the people in the white suits were that morning, but I realized that the pollution must be worse than my landlord and Louie made it seem if these folks were wearing protective gear to this extent. I continued wide-eyed along my way up to Smith Street for a croissant and coffee from one of the nearby fancy cafés, pulling my thin cardigan around me. I held my breath until I got to the corner and then I couldn't hold it anymore.

I lasted only a year at Smith and Fifth. The single streetlamp on the block burned out after a few months, and despite my weekly visits to the neighborhood councilwoman's office further up on Smith Street, our street remained pitch black throughout the winter months. The upstairs neighbor fractured his back at work a few weeks after I moved in and recuperated at his parents' house for the next few months, so the single mother and her daughter who lived next door were suddenly my only neighbors on the very dark block. Louie the junkie got married, and he and his wife grew increasingly hostile, refusing to leave when we asked them to, getting into fistfights with each other that left blood on the sidewalk. In early spring a man with a heavy, scraggly beard and red backpack appeared and took up residence in the corner between the bakery and my apartment. He stood like a block of wood, staring deeply into the empty lot, all night. During the day, when the bakery was operational, he sat on a milk crate next to a bodega around the corner, staring into a bush. He gave me the creeps in a way that Louie and his wife and the other visitors never did, and by later that spring I had called the cops so often that they stopped sending patrol cars. Instead, if it was before midnight, a man with a shaved head in a blue uniform would pedal his ten-speed up the block and weakly try to move people along.

Then my boyfriend became my husband, and with our combined paychecks we crossed to the other side of the elevated subway tracks. Our new apartment was two blocks away in a real brownstone, and our old Italian landlady spent most of her time mopping and bleaching the hallway floors. The first time I saw our squat green house after moving out, I was on the train as it swooped in an arc over the giant open lot. The house looked small and sagging, and I noticed new pink survey ribbons dotting the empty lot's curling barbed wire.

The Columbia University students I saw walking around in the moonsuits produced a project based on their findings at Public Place. Working with a professor from the School of Public Health, they developed a blueprint for an urban Vertical Farm to be built on the lot, in concert with open walking paths and a community center. In October 2007, the city opened a Request for Proposals from builders regarding the Public Place site. The RFP requested that proposals include suggestions for brownfield contamination remediation, as well as ideas for between four hundred and one thousand units of housing, commercial units, and a combination of affordable and luxury rentals. Apparently the Vertical Farm didn't interest them. David Walentas, the developer who had single-handedly turned the nearby Brooklyn neighborhood of DUMBO from a slum into a luxury artsy enclave in the previous few years, was named an early favorite. While the winning developer would be saddled with the responsibility of cleaning up the toxic waste, the tax breaks involved are incredibly lucrative.

The developers and the community groups fought for years, and the old lot just sat, waiting. Meanwhile, more studies were requested, and more moonsuited folks descended on the area. More attention was drawn to the Gowanus, and soon the word *Superfund* could be heard at community meetings. Some folks worried that the developers would take too many shortcuts in cleaning up the waste, or that even more toxins would be released during the construction process, and that the Superfund program was the only way to be sure the land was remediated correctly and safely. But many feared the Superfund designation, more than they seemed to fear the pollution itself. Many of the locals, like my landlord, who were hoping the new construction would be a boon for their property values and nearby stores, cautioned that the Superfund process would take too long, be too complicated, that once government became involved there would be no development, just a slow clean-up and tax dollars down the drain. The pollution isn't really even that bad, they'd say at meetings. *Think of the money!* The idea that illness due to exposure to the chemical soup could preclude their—or their children's—enjoyment of the windfall didn't seem to enter into the equation.

But as more studies were done and it became clear just how much of the shoreline of the Venice of Brooklyn was compromised by the area's industrial past, concern grew. The EPA estimated that clean-up of the 1.8-mile canal would take more than a decade and cost in the ballpark of $500 million. On March 2, 2010, the entire Gowanus Canal was inducted into the Superfund program, and the high-flying hopes for shiny new rows of apartment buildings and restaurants caught and stuck in the downward swirl of the canal's shit-brown sludge.

Sometimes, walking around my old neighborhood, I still see my former landlord. I don't mention Superfund. I don't mention the developers who fought like so many sharks in the murky water to take over the toxic waste, or the way the promise of the high-rises went poof. I don't mention that biology students from the New York City College of Technology detected gonorrhea in a drop of water from the Gowanus Canal. Instead, we talk about the weather. During these conversations, I notice that his shoulders seem perpetually slumped, like flat tires, and he has the look of someone whose sure thing fell through.

And every so often, in newspaper photos, I see the pea-green face of the old squat house just beyond the overgrown weeds and barbed-wire fence, peeking through like a distant artifact, someone else's home now. The house also seems to be slumping even more than it was when I lived there, as if at any moment it could give its final groan before collapsing in on itself and getting caught in the quicksand suck, joining the chemical sludge, the glittering black pearls, and the swirling dreams of real estate riches, sinking back down into the dark, moist earth, like so many rotting vegetables.

Corner Garden

Dara Ross

The D train rumbles underground as heat rising in waves from subway grates makes corner-store flowers wilt. But here, before the bulldozers, in the name of affordable housing, came into the garden that we made of this vacant lot, summer sun made budding flowers bloom.

I miss our garden as I walk down the block. I pass Julio, Felix, and Manny sitting on milk crates in front of the bodega, talking their usual talk.

"Boy, this heat sure is a bitch."

"Shit! This ain't no heat! Last week when I was down in San Antonio visiting Magdalena in school it was at least 110 degrees in the shade. I was out there just cryin' sweat. It was so hot that even them rattlesnakes out there was sweatin'! Shoot, I'd take this ol' city heat over some ol' desert dry heat any day, Jack."

The three were retired civil service workers living on pensions and Social Security checks. Julio used to drive a bus, Felix was a hospital janitor, and Manny was a postal worker. Due to limited economic means, but also maybe due to their city roots, they did not follow the migration of senior citizens to Florida. Instead, they spent nearly every day sitting in front of the bodega. When I came out of the subway, on my way home from work, I passed them sitting on these crates like magpies on a perch. They were always engrossed in conversation. Unlike Loudtalking Zora, the three observed the goings-on in the neighborhood but rarely made comments on them. Usually if they were not talking about the weather, or if Julio was not dominating the conversation with tales of his extensive travels, they talked about either the Knicks or the Yankees.

I overhear them talking about the same things I overheard them talking about in the garden. They talked about going to the Yankees' opener as Felix planted neat rows of radish seeds in freshly turned soil. They talked about how hot it was while Manny carefully transplanted indigo petunia seedlings that he had been nurturing for weeks into carefully hoed soil. They talked about how the Knicks were "going to go all the way!" this year while Julio planted hyacinth bulbs. Now they sit and talk their talk in front of the bodega.

On hot summer nights like this, everyone has taken to sitting. We sit on our front stoops watching traffic go by, yearning for the open spaces and the cool quiet places of our garden. Things are not the same around here. Mornings are no longer getting up early to pick mint leaves for tea before grinding off to work. Evenings are no longer for watering Universal pansies and impatiens with your grandmother's rusted tin watering can. Weekends are no longer for making preserves out of apples and rosemary sprigs from the garden. In late May, the cackling songs of migratory birds no longer drown out the sounds of boom boxes and police sirens.

Everyone who lives in these apartments above valued the open space of the garden. We loved the idea of having our own land, where we planted what we wanted and reaped what we sowed in order to woo a lover or to remember a lover past, to make scented bath oils and soap, or to moonshine strawberry wine. Even before there was a community garden, lots of folks filled their fire escapes with window boxes that spilled marigolds, poppies, petunias, pansies, and impatiens over the iron railings. There was hardly a windowsill or a fire escape without some flowers growing on it. Every corner where there could be a planter filled with tall tulips or a hanging basket of ferns, there was one. The people in this neighborhood love to grow things. But only after the garden was begun were so many people able to fulfill their dreams of farming fresh vegetables or raising patches of their favorite flowers.

Once we realized that our garden had become endangered and that we might lose its comforts, once we found out that the city was looking to demolish our little plot of nature, we fought as hard as we could. Eventually we failed as political activists to claim what we had worked so hard to grow as community activists. They said the garden had to go. That the land belonged to the city. The land was never ours to grow anything on. But we had made it ours. We had each participated in cleaning up waste and cultivating beauty. Even though Julio, Manny, Felix, and Loudtalking Zora are still here, R.C. and Dee have moved away, Wole has gone to college, and Hattie and Pearl have passed on. Some people, like Madeline and Evie, never bother any more to come outside. Like me they just get up, go to work, and come home. I miss them all deeply as I lie on my bed sweating and mourning in my hot, tiny studio apartment. I can hear Loudtalking Zora sitting on the stoop under my window, talking to voices I don't recognize.

"Girl, 'member when it was hot like this and we usta watch movies in the garden?"

"I sho' do girl."

"MmmHmmm. That sho' wuz nice."

"Girl, Julio usta set up that ol' projector—Julio! Julio!" Zora yells at Julio across the street. He was probably still sitting in front of the bodega. "Where did you ever find that ol' clunky projector anyways?"

"I find that beautiful piece of junk for twenty dollars at a flea market just outside of Sweetwater, Texas." Julio yells back. "I was on the way back from visiting Maggie and the grands in San Antonio and as I'm driving back home I see this flea market on the side of the road." His voice gets closer. "This old Indian guy is selling movie projectors, that's all he's selling is movie projectors on the side of the interstate. He had at least forty of them out there and as I was passing by I thought to myself, 'Who in the world would want one of those old, beat up, rusty projectors?' And then I remembered when I was a little kid in Santo Domingo we had only one TV in our village that belonged to our church. Whenever something especial happened on TV, the pastor would roll the TV out in the churchyard and the whole village came outside during the nighttime to watch it. I remember that there was something about watching TV outside under the open sky in Santo Domingo that was mesmerizing . . ."

"Girl, you should have seen it though," says Loudtalking Zora. "People dat ain't never planted nothing never in that garden would be up in there spread out in their lawn chairs, ready to watch them a movie."

"Oh go on, girl!"

"I sho' could go on for watching a movie like that on a hot-ass night like this."

Me too. I am relieved to know that I am not the only one who mourns for movies in the garden when it is hot like this.

The experience of watching movies in the garden, outside, under the night sky, chilling in your favorite chair, transcended all other moviegoing experiences. Julio's flea-market projector whizzed and whirled and spat out all of these images onto a white sheet that he had taped up onto the brick wall of the tenement houses that bordered the garden. Black and white 1930s Hollywood musicals. Bugs Bunny cartoons. Independent art-house films. Any movie we had the privilege of finding a reel for got four stars due to its accompaniment of cooling breezes and the scents of lavender herb and yellow honeysuckle mingled with white China roses. When the evening turned toward midnight and got very chilly, Pearl would wrap sleeping children with any number of brightly colored afghans that she had crocheted with her dark and tiny wrinkling hands.

Our garden was the center of it all. It was a place for people who were not fortunate enough to have their own private plot of back yard to grow vegetables or flowers. It was a place where one could share in the joys and the rewards of nurturing and growing things. When I first moved to this block, the vacant lot on the corner looked as dim and as dirty as any vacant lot in this city. None of the old timers could remember how long this wasted space had sat and sat. No one could even recall what had stood there before the junkies and drug pushers took over.

"It was R.C. that started it all, you know." I hear Zora below. This time she is talking to Alma.

"You know I always usta pass by that lot every day on ma way to work and think to ma self, 'Look at this nasty-ass lot. What a waste. Now why don't somebody do something about it?' And one day on ma way to work, I see R.C. in there with a box of garbage bags just picking up all the garbage. And I say to him, 'Boy, what are you doing in there?' and he say, 'Morning Ms. Zora,' you know all polite-like, you 'member how he usta talk with his fine-ass self, 'Morning Ms. Zora, I'm just clearing a little space here, I thought I would plant me some vegetables or maybe some flowers.' Just simple as that! 'I thought I would plant me some vegetables or maybe some flowers.' By time I got home from work that day, can you believe that R.C. was still on out there picking up garbage?"

"Are you for real?"

"You know I am, girl. But lemme tell you—not only was R.C. out there cleaning up that nasty-ass mess but so was JoJo, Liliana, Lester, Calvin, Sherrie, NayNay, and even that lazybones Willie was out there helping R.C. And that's how it went every day. Everyone helped out a little 'til they got that whole lot cleared up."

"Shoot girl, now I know that was a lot of work."

"Sho' nuff it was."

R.C. was a beautiful man. He was tall and brown and lean, and he had a mischievous smile. His back was strong and so was his heart. He always had time to help anyone with anything, especially when it came to the garden. He had helped just about everyone, at one time or another, with their planting, whether it was clearing out weeds or turning stubborn soil. This man was disciplined, determined, and dedicated enough to start the process of creation. And he stuck with it right until the very end. Right until the bulldozers came and plowed over what he had worked so hard, what we all had worked so hard, to cultivate. That garden was so many things to so many people. Besides a place to grow things and an outdoor movie theater in the middle of a community that was too poor to have a multiplex, that garden was also a reception hall, a meeting place where neighbors swapped recipes and seedlings, a place where the little kids in our neighborhood learned about the cycle of butterflies, a place to sit and contemplate, a place to appreciate the quiet of sleeping ladybugs.

Of course, if you were tired of quiet and contemplation or if you were bored and wanted to dish some dirt, Zora and her girlfriends turned the garden into the grapevine. At about 6:30 in the evening, after they changed out of their suits and sneakers and into their garden overalls and sweatsuits, Zora, Alma, Lizettte, and Lenicha would stand about with garden tools that I never saw them use and talk the garden gossip. They didn't do much planting, but they always had the shiniest garden tools and the "real low down" on who was doing what, where they were doing it, and with whom. After the scoop was laid somebody would say, "OoooWeeee girl, now you know that she didn't need to go there!"

"Who you telling? Shoot, I know."

"But she did though, and that just goes to show you what kind of fast-ass that girl really is. You know what I'm saying?"

"I sure do, girl. You know I do. Lordy. Lordy. Lordy."

If you weren't interested in what the gossips had heard and if you didn't want to do any planting, if you just wanted to be lonely, there was always a quiet spot near the corner of the garden where Julio had put a little cast iron loveseat that he picked up at a flea market outside Buffalo on his way back from visiting a cousin in Quebec. Julio did his best to get around. The bench was lopsided, a bit rusty, and vines of ivy had started to wrap themselves around its frame, but it was here that you could sit alone among the mis-mosh of mismatched patches of flowers and vegetables that we had randomly planted and be mesmerized by all of the different shapes, sizes, colors, and fragrances. In our garden, everyone had space enough to plant whatever they favored for whatever reason they could think of. This is what gave the garden flavor. And it did, like the people who cultivated it, have a flavor all its own.

The blue of Evie's salvia plants grew next to Madeline's santolina grasses. Wole's crimson snapdragons lay at the feet of Mr. Nate's strawberry plants. Pearl's huge leafy-green Royal Standard hosta plants towered in the shade over Manny's plot of petunias. Felix's burnt-orange African daisies bordered Hattie's Black Ball cornflowers, and Julio's fragrant lavender stalks grew next to Delano's rosebush.

My friend Madeline from Nevis, a six-mile-wide island in the Caribbean, lived down the hall from me. She was working on a bachelor's degree in chemistry and was planning to be a junior-high-school teacher. Her pastime away from school was planting herbs in the garden so that she could make her own beauty products. She dreamed of having her own line of products one day. She was the reason our garden smelled the way it did. The things she planted made the garden smell like a French parfumerie. Her plot was filled mostly with sweet-smelling herbs like elderflower root, meadowsweet, rosemary, and chamomile. She also planted a few roses, and she even grew chocolate-scented cosmos. She amazed me with what she could make once she harvested her crops. Even before the garden existed, Madeline grew herbs on her windowsills. On any Saturday night we sat in her kitchen making face toner, bath oil, and my favorite, bar soap.

Madeline showed me how to make fragranced bar soap out of bars of unscented glycerin soap that she bought at the drugstore and melted down in a double boiler. She would add a tablespoon of fine oatmeal and then stir in some chopped herbs that she would have picked from the garden minutes before preparing the soap. I liked to mix a batch using fennel with chamomile and maybe some mint, but Madeline would go all out, mixing bits of petals from a Fragrant Cloud rose with elderflower root, rosemary, meadowsweet, calamint, grated orange rinds, and crushed mango seeds. "In Nevis there are so many different smells that blow across the island. The island is only about six miles across at its widest point. On a clear spring evening a strong breeze will carry the scent of calla lilies and ripe passion fruit clear across the island. Right into my bedroom window. I want something that smells like the islands," she said, while pouring the herbal, oatmeal, and glycerine mixture into molds lined with wax paper. "Something to remind me of home."

Mr. Nate, who came from a long line of southern backwoods moonshiners and wine distributors, planted strawberries for the purpose of making illicit strawberry wine. All of us gardeners couldn't wait until late September once the wine had fermented in Mr. Nate's basement. Once the wine was fermented, Mr. Nate would hand out bottles and bottles of sweet strawberry wine. "My poppa and my great-grand before him used to moonshine out of the backwoods of South Carolina," Mr. Nate would say while tending to his strawberry patches. "My poppa used to make these sweet wines for my momma and her sisters. Boy, did she love her some of poppa's strawberry wine. He knew how to make it too. Real sweet-like so you hardly knew that you was drinking wine. You just thought that you was drinking strawberry juice." Mr. Nate's momma sure knew where it was at, because on a hot summer night there was nothing like it, getting drunk off strawberry wine, laughing and chatting with your girlfriends in the dark cool of the garden.

Hattie always planted Black Ball cornflowers because they reminded her of South Carolina. She had moved up here to be closer to her children after her husband, to whom she had been married for sixty-two years, had died. Hattie was in her late eighties but still found enough strength to not only get up in the mornings, but also to get down on her blackened knees to tend to her cornflowers, "Strangest little flowers you've ever seen, ain't they?" she once asked out of the blue across our plots, as she dug down in the dirt pulling out weeds as I watered my string bean stalks. "I was sixteen years old when I married Jeremiah, I've known him all of my life, it seems. And all of my life, ever since he first started courtin' me when I was just fourteen, barely able to wear stockings, Jeremiah brought me cornflowers."

There are a lot of older folks in our neighborhood. They loved to plant flowers that remind them of days past and vegetables that they could make their famous recipes with. I would plant whatever seeds I could get on sale at the five-and-dime store. I had bad luck planting, because nothing I ever grew returned the following year. Only the weeds returned. So every year I would start my planting over from scratch. One year I grew a row of snapdragons and some lettuce. Another year I grew some sweet pea vines on a stick and some string beans that I usually ate raw off the vine, thereby never harvesting enough to cook, much unlike Liliana who lived across the hall from me.

Liliana was from Belize, and she knew how to plant vegetables. Her plot took up more space in the garden than most of ours, but no one minded because everything that Liliana planted grew. And even better, everything that Liliana planted she had a special recipe for. And even more importantly than that, Liliana loved to cook for others. She grew tomatoes and canned her own tomato sauce; and Lord knows how she did it in the middle of Brooklyn, but she grew up sweet corn that she used in cornbread along with a little sprinkling of dill. She also grew cabbage and eggplants. I spent many afternoons in Ms. Lilie's kitchen eating fried eggplants or stewed cabbage with tomatoes.

"This is how we did it in Belize. Back home, we grew whatever we could grow and we ate what we grew," she said one day while she showed me how to make spaghetti sauce. "When I was a little girl people used to tell me about how in the United States only a few people grew what they ate, and that most people bought what other people grew for them to eat, and I did not understand."

"I love this garden. It reminds me of home. Everyone grows something and everyone shares what they grow. You see this fresh rosemary, Madeline gave it to me and I gave her some corn. I am going to make some strawberry ice cream with some of Mr. Nate's strawberries tomorrow night. My little granddaughter Maria is coming to visit this weekend. Would you like to help make some homemade ice cream?"

The day that R.C. and the rest of us had finished clearing away all of the garbage and turned the soil, Liliana planted a bare-rooted apple tree. The tree grew Lane's Prince Albert apples and they were red and sweet. Liliana made apple jam preserves out of them. Once Madeline and I made some apple jelly with rosemary to use as a glaze for cooking. Liliana tended to that tree all winter, pruning its branches and telling the tree stories during the months when only the hardcore gardeners were around. That tree should have been bearing fruit for jams, jellies, pies, and breads for years to come. When the city came to clear away the garden we all stood and watched. It took them all day. And all day we stood and watched, but it wasn't until they uprooted Liliana's apple tree that we started to weep.

R.C., who started the garden, planted the most ebullient flowers. "I was twelve years old when my grandmother taught me the importance of giving ladybugs, grasshoppers, and daddy longlegs free rein in your garden," he told me once. His matrilineal heritage was one of growing things, and he knew how to grow the kind of flowers that if cut one week, would be in bloom again the next week. He always had a fresh supply of euphorbias, forget-me-nots, love-in-a-mist, sweet peas, nicotianas, baby's breath, and primroses for Dee, with whom he was in love. She lived around the corner and went to Hunter College, and when she didn't wear cowrie shells or beads in her long braids she would adorn them with the flowers that R.C. grew for her.

After a couple of weeks of being bombarded by sweet bouquets of love-in-a-mist and roses from R.C., Dee started hanging out in the garden and tending to his flowers. She wanted to grow some okra and a few tomato plants of her own—she was a big okra eater. She loved fried okra, stewed okra and tomatoes, and especially gumbo. R.C. graciously cleared some space for her in his plot. By late that summer, Dee's okra plants and tomatoes were growing up among R.C.'s cutting plants in a crush, and the following spring they were married in the garden.

We all gathered around them as they stood under the trestle that Julio bought at a flea market outside of New Orleans. It was covered in a sweet tangle of climbing roses and raspberry vines. R.C. and Dee quoted passages from the Song of Solomon during their outdoor ceremony, which took place in the garden at dusk.

> I am the Rose of Sharon and the Lily of the Valleys.
> As the lily among thorns, so is my love among the daughters.
> As the apple tree among the trees of the wood, so is my beloved among the sons.
> I sat down under his shadow with great delight, and his fruit was sweet to my taste.
> He brought me to the banqueting house and his banner over me was love.
>
> Awake, O North wind; and come, thou South;
> Blow upon my garden, that the spices thereof may flow out.
> Let my beloved come into his garden, and eat his pleasant fruits.

For wedding favors they gave out herb seedlings of lavender, dill, and sage to give thanks to everyone who participated in celebrating their marriage. The following morning, those of us who weren't hung over from Mr. Nate's strawberry wine were planting our seedlings.

As part of R.C. and Dee's wedding festivities, they planted another tree in the garden. This time it was a nectarine tree, and it grew up just as fast as the apple tree that Liliana had planted five years earlier, prompting the gardeners to believe that there was a river flowing deep underneath the garden watering the roots of everything that we planted. R.C. and Dee maintained that the plants in the garden grew out of our overwhelming sense of love and commitment to the garden. He said that his grandmother always told him that the plants knew when you loved them. And that there was nothing that grew up faster than a loved plant.

R.C. and Dee's wedding day was the height of the productivity in our garden. Smack in the middle of the seedier side of Brooklyn, we had paradise. In the spring and the summertime I could not wait to get home. Doing some work in the garden was a way to unwind from a stressful workday spent cooped up in an environmentally controlled atmosphere, especially on days when maybe for hours straight I would breathe only the same recycled air. I would rush to change into clothes destined to get the kind of dirty that did not wash right out and tend to my little plot of garden. By six o'clock in the evening, most of the retirees had finished their garden work. By the time I arrived they would be sitting in the shade telling stories, and soon I would be on my knees in the moist dirt listening to Mr. Nate tell stories about moonshine in the South or Liliana talking of her home in Belize.

Loudtalking Zora must have gone inside or moved from under my window. I grow sleepy. I can hear Hattie, Mr. Nate, and Pearl sitting in the garden talking about home and chatting away about sharecropping plantations and southern locations. I finally drift off to sleep and dream of their soft voices and then the sounds of crickets and the quiet of nesting birds. I smell the faint traces of purple hyacinths mingling with angelica leaves.

A Land Ethic for the City

William Kornblum

Students who take the risk of registering in my undergraduate environmental sociology course at Queens College will encounter the following poem in the first five minutes of the first day of class. It's not one of Robert Frost's best-known works, but it serves extremely well to get students thinking concretely about our heavy footprints on the earth. The poem begins to lead us toward the goal of the course, which is to use some of the classics in American environmental literature to arrive, as dwellers in the nation's most urban place, at a personal land ethic for the city.

A Brook in the City

The farmhouse lingers, though averse to square
With the new city street it has to wear
A number in. But what about the brook
That held the house as in an elbow-crook?
I ask as one who knew the brook, its strength
And impulse, having dipped a finger length
And made it leap my knuckle, having tossed
A flower to try its currents where they crossed.
The meadow grass could be cemented down
From growing under pavements of a town;
The apple trees be sent to hearth-stone flame.
Is water wood to serve a brook the same?

How else dispose of an immortal force
No longer needed? Staunch it at its source
With cinder loads dumped down? The brook was thrown
Deep in a sewer dungeon under stone
In fetid darkness still to live and run—
And all for nothing it had ever done
Except forget to go in fear perhaps.
No one would know except for ancient maps
That such a brook ran water. But I wonder
If from its being kept forever under,
The thoughts may not have risen that so keep
This new-built city from both work and sleep.

A bit confused over why a sociologist is handing them a poem, the students will avoid eye contact with me; I might call on them to offer some ideas of what the poem is about. But we'll read the poem together and do a little parsing of its possible meanings. I'll ask if they encountered, as I did in my Flushing childhood, any of the remaining old farmhouses that once lingered in many old Queens neighborhoods. For some awkward moments we'll also try to grasp the poet's spiritual notion that once cast down into a sewer trunk line, the brook can somehow deprive us of work and sleep. But before the students can rebel or simply begin to nod off, I'll ask them to come outside with me for a short field trip to see for themselves what Frost is getting at. Some images I've snatched from Google Earth and elsewhere can help the reader simulate that field trip.

A Campus at the Crossroads

To shift our thinking of the campus as just another urban place in Eastern Queens—served by a number of bus lines, with no subways and packed parking lots—the class follows me out of Powdermaker Hall and stops to look west over the city. From the terrace between the library and the science building we gain a splendid view over much of Queens and Manhattan. "Why is the land here so elevated? How did that happen?" I ask. Invariably at least one of the students will mention something about "the glaciers." That dredged memory of scientific fact leads to a brief discussion of how, after the last ice age, something like ten to twelve thousand years ago, the receding ice left a moraine, a hill of rock and stones and brought some elevation to a flat coastal plain that slopes gradually toward the edge of the continental shelf. We are starting to think of the landscape as the result of great forces working over time, beyond human control, forces far greater even than the impact of the city-building road scrapers and cement trucks. But we also agree that it's not wise to minimize the impact of our own activities on the landscape, either. We walk down the campus to the intersection of the Long Island Expressway (I-495) and Kissena Boulevard, the main street leading to the Queens College campus.

We are walking downhill, past the venerable Gino's Pizza restaurant and then to the crossroads at the foot of the campus. "What happens when it rains or snows and the water runs off the streets? Where does the water go?" I ask. This is not a question the students have been thinking much about. It seems at once obvious and obscure. The water, after all, goes into the sewers we readily see along the sidewalk curbs. I ask the questions again and the same student who remembered the glaciers volunteers that "eventually the water must run into the ocean." Eventually, but after it enters the sewers, where does the fresh water go? Where does it meet with salt water?

This question draws attention to the contours of the land we are walking on as it slopes not toward the ocean, but toward the western edge of Long Island Sound and the beginning of the East River (as shown in Figure 1), where two great bridges, the Throgs Neck and the Whitestone, cross that transitional section of the Sound. Over the semester, as we read Thoreau, Rachael Carson, and Aldo Leopold, we will keep trying to connect big ideas of environmentalism to the local specifics of our estuary, and to the impacts of what happens after the water sinks into the sewers—to problems, for example, of groundwater runoff and its impacts on the our vast urbanized estuary.

We stop on the overpass above the roaring Long Island Expressway. Figure 2, a close-up from the same photo as Figure 1, shows in the center where we gather, but it misses the discomfort we feel. The students are confused. They seem to be wondering why we are hesitating here in such an uninviting place. There's no danger; the sidewalk is wide and the railings high and strong, but the cars and trucks rush under our feet at sixty miles an hour. It's not all that easy to hear their teacher, so the students gather closer. "I know this is not the most pleasant place to stop and talk," I admit, "but we're making academic history here. How many other teachers try to hold part of a class over a piece of the interstate highway system? It is such a common scene in urban America, I thought we might try to enjoy it as part of our nature walk."

Figure 1

Figure 2

I point to the four corners where there are two large gasoline stations, a Dunkin Donuts shop, and on the fourth corner the concert and cultural center of the Queens College campus. "What do land uses at three of these four corners have in common?" The students seem perplexed. What am I asking? Two gas stations and a donut shop, and the concert hall: none are distinguishable from the aerial photo, but they stand out unmistakably from the ground level. "Well, think about the gas stations and the donut shop. If you think about what they are really about, you'll see what they have in common." Silence and foot shuffling. "Well, what do gasoline and donuts have in common?" I see some glimmerings of ideas in their eyes, but it's the first day of class and no one wants to hazard a possibly embarrassing guess. "What about energy? Aren't the gas stations and the donut shop about energy?" A few of the students groan. Again, the question seems so obvious now that the answer is dangling before them. "We think of fuel for the cars but don't equate it with fuel for the human body, but how different are they?" "Fuel for the cars and fuel for the body," one student quips, and I am thankful for that help. But if three of the four corners are devoted to energy and the basic biological needs, the remaining corner, the Colden Auditorium of Queens College, the large white building at the bottom left of Figure 2, is a space, the students conclude, dedicated to the higher aspirations of our species.

Before leaving the overpass I ask the students to look down at the expressway again, not to the hurtling traffic but to the sumacs and weeds clinging to every patch of earth along the roadway. We agree that each is a green tribute to resilience of life on this urban motorscape, but the students do not hide their sense of relief as we continue downhill toward Kissena Park.

It's relatively easy to see in Figure 1 that Kissena Park and its adjoining golf course are an essential part of a green corridor that channels the flow of water along a natural drainage basin. The water flows in sewage trunks and over surfaces that include the wetlands of Flushing Meadow Park in the old World's Fair Grounds, and finally into the East River at Flushing Bay, to the west of the Whitestone Bridge. It's far more difficult to understand where the park fits in the region's estuarine system while we are walking along the congested streets between the campus and the park. We can imagine the water flowing along the concrete curbs and into open sewers, but it takes the kind of overview a map or aerial photo such as Figure 3 offers to get the real lay of the land. On the other hand, unless we leave the classroom to take a walk through the landscape, the discussion of where the campus and its surrounding neighborhoods fit in as part of an urban estuary remains rather abstract.

Figure 3

Out of twenty-five students only one has ever visited Kissena Park, and that because he grew up in its immediate neighborhood, and for him it was the local play area. Most of the other students come from other neighborhoods in Queens, where there were other parks nearby. Although the park is only about five blocks from the campus, few students have reason to go there. Like the lone student who knew the park, I too grew up within walking distance of its lake and athletic fields. Kissena Park was a significant place in my own experience of childhood, but understanding of the role it plays in the regional watershed, and the insights I gained about the park from reading Frost's poem, are far more recent. I do know, however, that the ancient creek that drained this now-ruined wetland remains in the park and is submerged in most places, "thrown deep in a sewer dungeon under stone." Figure 3 shows a path that leads past the ballfields and toward the small lake in the center of the park.

It's clear (once I point it out) that the cement path running toward Kissena Pond is bordered on our left by a small "forest" of stately deciduous trees, and by a much-disturbed wetland to our right. Below the surface of the path there is running water, and its presence is given away in the boggy places where there are stands of willows and tall Phragmites grass. A bit farther I take the students off the path to a place where we can actually see the water where the stream is not buried. We trace this mostly hidden stream to the concrete shore of the lake itself (evident in Figure 3), and I point out that if we look behind us from that vantage, we can clearly see the way the land slopes down from the campus (visible in the near distance) and from the park to form a narrow valley in which the water must run toward Flushing Meadows, the site of the city's World's Fairs. In this corridor of former wetlands and preserved patches of open space I can also show traces of animal habitat—rabbits, raccoons, shore birds, upland birds—in addition to the usual squirrels and pigeons. It's almost time for me to dismiss the group so that the students have time to get back to the campus for their next classes, but for a few moments I try to get them to come up with explanations of why it matters whether we know which way the water drains off the land, and where there are some remnants of older, nonurban ecosystems. A few of the students' comments indicate that they are getting the point of our "field trip." Most are still looking unimpressed and a bit confused. I leave the class with the assignment that they write a brief essay on why it matters if we know where the water drains from the surface of the land and what it carries as it flows.

Over the course of the semester we will read a few classics in American environmental thought—*Walden*, *A Sand County Almanac*, *The Sea Around Us*, and *Silent Spring* notable among them. In most of the readings the authors draw on vast scientific and personal knowledge of specific ecosystems to make pleas for environmental preservation and restoration. But few of the classic readings deal with nature in cities. In consequence, many of the additional reading assignments over the semester will draw on ideas from the classics as they are applied to the urban environment.

A Land Ethic for the Interstate

Somewhat later in the semester, especially when we read *A Sand County Almanac*, we'll think specifically about Aldo Leopold's deceptively simple plea that we each take on "the oldest task in human history; to live on a piece of land without spoiling it." Of course each class will inevitably deal with one or another aspect of this plea, but for city dwellers there are special twists and turns in the discussion that often arise from debates about taxes, and more specifically, about the use of public funds to protect environmental quality.

A few miles from the intersection of 1–495 and Kissena Boulevard, where the students and I stood above the rushing traffic, there is another intersection where from 2000 to 2006 motorists, including many students and faculty at Queens College, spent miserable amounts of time in traffic jams due to a major construction project. This is the intersection of I-495 and the Cross Island Parkway. It is situated in a portion of the old wetlands that drains toward the Long Island Sound in eastern Queens, in the vicinity of the Throgs Neck Bridge. Most people who suffered through the years of traffic congestion due to road construction understood that the delays were due to the creation of a modern cloverleaf intersection, a significant modernization of the roadway system, but only the most environmentally sophisticated were aware that the project also included some rather innovative and expensive environmental restoration in and around the new intersection, as indicated in Figure 4 and in the passage quoted below from official descriptions of the project.

> Twelve acres of Alley Pond Park will be restored and/or reintegrated via the elimination of the two existing highway loop ramps, constructed forty years ago, which separated the park into segments.
>
> The sediment-filled Alley Pond, located within the southeast quadrant of the LIE/CIP Interchange, will be reconstructed . . . and restored to an open water body that includes native emergent marsh and bio-remediation plantings. Reconfigured interpretive trails, including a scenic overlook, will be provided to improve access to the restored Alley Pond.
>
> Extensive environmental mitigation work will be performed on Alley Pond Park, including new landscape plantings and reforestation with native species that will enhance the existing vegetation. A landscaped earth berm will be fashioned in the southeast quadrant to help screen the park area from the LIE. To eliminate the substantial current drainage and erosion problems, improvements will be provided at a number of locations including areas along the Tulip Tree Trail and Alley Pond south of West Alley Road. Two detention ponds will be created within the northeast quadrant of the LIE/CIP Interchange to improve the quality of storm runoff to

Alley Creek. These ponds will be vegetated with bio-remediation wetland plantings capable of removing pollutants from the storm runoff. Oil separator drainage structures will also help remove pollutants.

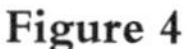
Figure 4

Alley Pond Park, far more than Kissena Park, is a popular site for environmental studies, largely due to the non-profit Alley Pond Environmental Center "dedicated to establishing an awareness, understanding and appreciation of the environment and the responsibilities associated with preserving the environment in an urban setting." My students had never visited the center, but class discussion of the interchange project, and discussion of terms like "bio-remediation," "a landscaped earth berm," "reforestation with native species," and "oil separator drainage" introduced them to a hopeful place in the heart of the region's highway system. At the same time, we had to recognize that environmental preservation and restoration can be expensive. The project's initial cost estimate was $112 million, and the Alley Pond interchange example raises a far more common problem: public expenditure on the environment is frequently invisible to the taxpaying public. But trying to develop a land ethic for city folk means, among other things, that we need to know how and why tax dollars are spent on stewardship of the land and water, and why these expenditures are taking on ever-greater urgency in an era of global warming.

Seas Around Us: Cities and Metropolitan Regions on a Sunken Coastline

The entire eastern seacoast of the United States from Cape Cod to the Florida Keys is essentially a flat coastal plain, with wetlands and beaches sloping gently into the ocean. This is also true of the entire Gulf Coast, and the entire coastline extending along the eastern coast of Mexico and South America. The west coast of both North and South America is mountainous, with small beaches carved out of ancient coastal canyons. Sociology courses do not usually delve too deeply into plate tectonics, but to understand the environmental politics and policy debates that arise along our shores and cities, we need background in the basics of coastal geomorphology, and this, in turn, requires an understanding about why the East Coast slopes into the sea while the western one is thrust skyward.

Geomorphologic differences between the two coasts took on far greater relevance after the severe storms of 2006. Stark images of drowned bodies in backyards and ships cast up on street corners made discussion of "the power of nature" far less abstract than ever and drove home the essential interdisciplinary nature of ecological knowledge. Hurricane Katrina revealed to Americans of all ages, and especially to college students, why we need to better understand the impact of coastal urbanization on coastal ecosystems. Everyone who listened to accounts of why the storm was so damaging heard about the influence of diminishing wetlands and barrier islands along the Louisiana and Mississippi Gulf Coasts. When I show my students the photos in Figures 5 and 6, however, they don't adequately make the connections between the coastal zone of the storm-devastated Gulf and their own coastal zone, so these figures can help teach the basics of coastal-zone literacy to nonbiology students.

Where does all the sand on the beaches come from? What causes the sand to pile up in the form of barrier islands like Jones Beach or Fire Island (or the remnants of barrier islands we see in the city, like the Rockaways and Coney Island)? How does the action of the tides help

Figure 5

define different zones of the wetland ecosystem? Why are tidal lagoons like Great South Bay and Jamaica Bay in our region (or the great-but-endangered Chesapeake) among the most productive ecosystems on the globe? What part do the marsh grasses play in this productivity? These and many related questions could occupy much of the semester, but for nonbiology majors the essential processes of coastal-zone dynamics can be summarized in a class session or two. And they must be covered, not only as basic environmental literacy for those of us living on the sinking coastline, but as a prerequisite to any study of the "political economy of sand" (Figure 7).

Every year the Army Corps of Engineers awards millions of dollars in construction contracts to private companies that are in the business of pumping sand onto local beaches. Again, these are tax dollars that the public does not usually see in action. The public wants the beaches preserved, but how many understand the costs involved, or, even more importantly, why those costs arise in the first place? Sand is transported by currents that run parallel to the shoreline. Wherever these currents are slowed by natural or man-made obstructions, the sand is deposited. Figure 6 shows this process in action at Coney Island.

Figure 6

The huge rock jetties along the beach are meant to diminish wave energy and preserve the sand from eroding currents. Sand accretes on the upstream side of the jetty, but the water rushes around the rocks and scours sand from the downstream side, creating the scalloped effect we see in the photo. The scouring effects are worst where the line of jetties ends, as is evident in the foreground of the photo. In consequence, once one jetty is built to preserve sand along a barrier beach, the entire beach must be "stabilized" by a line of such jetties, or major erosion and an unwanted breach may occur. Although the nation's barrier islands are highly unstable environments, federally subsidized insurance policies and other prodevelopment real estate measures have encouraged building on them. In consequence, the high-rise buildings that line the shore of Coney Island are becoming a ubiquitous feature of the coastline from New York to Florida and around the Gulf Coast. This development encourages further efforts to pump sand and build jetties, but as we saw with Hurricane Katrina, the damage caused by the 100-year storm is proportional to the level of development on the sinking shore.

Only 10 percent of the entire shoreline of the United States is in the public domain, which goes a long way toward explaining why there is so much urban development on barrier islands and on coastal landfills. This is a fact that shocks many Americans, especially students, and it requires some further explanation. From common law and ancient precedents, the shoreline below the mean high-water line is within the public domain, which means that anyone can walk along the beach even if beyond the tide line there are villas, condos, or other private developments. Access to beaches and barrier islands can be barred to the public, however, simply by imposing parking and trespassing restrictions, and we see this everywhere in the New York metropolitan region and elsewhere in the United States In consequence, any public parks and other public lands on the shoreline are most precious. An essential aspect of an urban land ethic for a coastal city such as New York is advocacy for public access to the beaches and the water. Study of Gateway National Park becomes a critical case study, at least for residents of our metropolitan region, for understanding the costs and benefits of winning public access to the water.

Gateway, a New Breed of Urban National Park

Out in the marshy center of Jamaica Bay, at the bottom of the city near the end of the endless A line, is the nation's (perhaps the world's) only wildlife refuge accessible by subway. The renowned Jamaica Bay Wildlife Refuge is a unit of the Gateway National Recreation Area. Formed around two freshwater ponds and bordering on the grassy hummocks and open waters of Jamaica Bay, this is a magical place where flights of ducks and other shore birds, including herons, egrets, and ibises, land and take off undisturbed below intercontinental jets on their way to and from nearby JFK International Airport. Far in the distance, over the wetlands and the expanse of Kings County, the Manhattan skyline is visible on the horizon. And none of this would be available to us had it not been for the heroic efforts of the legendary New York park keeper and horticulturalist Herbert Johnson. In the late 1960s Johnson took a bulldozer to the heaps of mud and sand that had been left on Broad Channel Island after construction of the A line across Jamaica Bay. Johnson was an avid birder and knew that if he could impound fresh water in the marshland, the pond would attract flights of migratory birds as they made their way up and down the Atlantic flyway during their seasonal migrations. With permission from his boss, the master builder and omnicommissioner Robert Moses, Johnson went ahead and built the berms that made the wildlife refuge possible. The new pond quickly became a favorite destination for thousands of birders throughout the region, and eventually, as Congress in the early 1970s developed plans to create national parks in some of the nation's metropolitan regions, the existence of the immensely popular refuge in Jamaica Bay was a powerful inducement to the creation of Gateway National Recreation Area, which along with its sister park in San Francisco, Golden Gate National Recreation Area, became one of the nation's first two urban national parks.

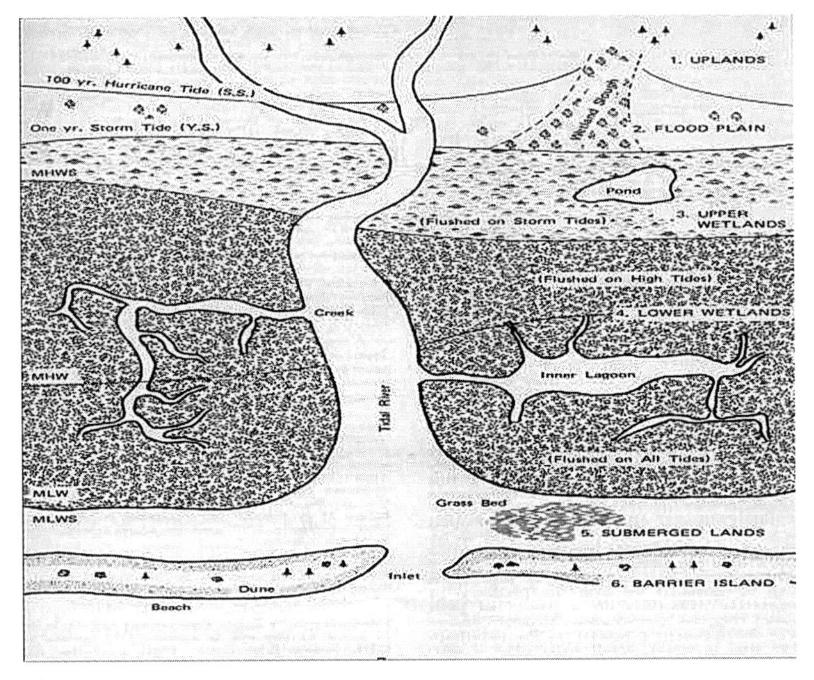

Figure 7

The story of Herb Johnson—a modest man, educated in the city schools—has a moral: environmentally active citizens we can make a difference in how deep our footprints are on the land. Making this kind of difference requires knowledge and intimate experience of our regional ecosystems. As soon as one looks closely at a park like Gateway, it becomes clear that we cannot assume that just because it exists it will always be lending its federal protection to the shoreline and wetlands of the city's outer harbor. One perennial problem the park faces is a dearth of spectacular geological features and heroic species.

Heroic *Limulus*

The great national parks of the West boast unique geological features such as Yosemite Valley and El Capitan, or the glaciers (while they last) of Glacier National Park in Montana, or the caldera of Yellowstone. Gateway has urban barrier beaches along the sunken shoreline of the Rockaways and Sandy Hook, and the marshes of Jamaica Bay. Visitors crowd the western parks to see bison and bald eagles and bears, but an urban national park like Gateway has none of these heroic species. But we do have *Limulus polyphemus*, the lowly and much-abused horseshoe crab. As a child I learned from the older kids in my neighborhood with whom I poked along the polluted edges of Long Island Sound or Flushing Bay that the horseshoe crab was a dangerous critter. It was hard to see in the murky water, they said, and if one stepped on its rigid tail it could pierce a foot. So we felt duty bound to kill it wherever we found it crawling in the marshes or on the beach. We also saw that fishermen harvested the slow-moving horseshoe crabs by the thousands, only to cut them up alive to bait their traps. In fact these animals are not crabs at all. They are arthropods, living descendents of the extinct trilobite; they are living fossils older than the dinosaurs, survivors of the mass extinction at the end of the Paleozoic Era more than 250 million years ago. *Limulus* is a heroic species here in the metropolitan region, and although all the environmental education we can muster will probably never raise its status to that of the grizzly bear or bald eagle, if we can begin to understand why *Limulus* is to be praised and preserved instead of subject to wanton destruction, we will have moved another step toward a land ethic for the city.

Adapting Leopold for Our Time and City

Conservationist Aldo Leopold framed his concept of the land ethic out of his dissatisfaction with conventional conservation ethics. He pointed out that we cannot be content to create national parks and other conservation areas when in our own backyards we pollute and despoil the land and waters. He said,

> The land ethic simply enlarges the boundaries of the community to include soils, waters, plants, and animals, or collectively: the land.
>
> This sounds simple: do we not already sing our love for and obligation to the land of the free and the home of the brave? Yes,

> but just what and whom do we love? Certainly not the soil, which we are sending helter-skelter downriver. Certainly not the waters, which we assume have no function except to turn turbines, float barges, and carry off sewage. Certainly not the plants, of which we exterminate whole communities without batting an eye. Certainly not the animals, of which we have already extirpated many of the largest and most beautiful species.

Leopold was writing primarily for an audience of landowners, range managers, farmers, people with some property. City folk may not own land, but as I have been trying to show in this essay, they have enormous stakes in the land and water that surrounds them. For urban audiences, the first responsibility in developing a land ethic is to become environmentally literate, that is, to know how our regional ecosystems work and what behaviors threaten their health and continued existence. The second critical aspect of a land ethic for urbanites is to become environmentally active as advocates for public access to preserved natural areas within and outside the cities. This does not mean that every citizen must become obsessed with the problems of global warming, to the exclusion of other issues involving social justice, for example, but it does mean that at a minimum we understand how to defend the land, as Leopold broadly defined that term, everywhere we find it.

Can Naturalists and Urbanists Find Happiness Together?

Phillip Lopate

Why does nature matter to New Yorkers? Or maybe the question to be asked is: Why should New York and other big cities matter to naturalists? And can naturalists and urbanists find happiness together? We might all agree that they *should* find happiness together, but not necessarily that they *can* find happiness together. They should because, from my perspective, the real hope for the future is to have both cities and wilderness, and to find a way to contain suburban sprawl, which is now eating up one acre per hour. The only way that is going to happen is by our not just protecting the environment, but by recognizing that dense cities have a very important role to play in protecting the environment. That may necessitate a rearrangement of some people's attitudes toward density—particularly the environmentalist attitude. If indeed, as some experts have told us, the most energy-efficient site in America is a Manhattan street, then density should be considered a good by environmentalists as well as by urbanists.

I want to take a step backward first and consider the problem in context: where all the resistance to cities is coming from. We have to remember that it's a rather American problem, or prejudice, this idea that cities are bad. There has been an anti-urban bias in American thought at least since Jefferson, who was particularly adamant in his mistrust of cities. He mistrusted them partly because he saw the archetypal citizen of the new Republic as a farmer, a yeoman—in any case, as someone who engaged in agriculture—and partly because he saw cities as giving too much power to the bankers and businessmen. He warned against the threat of cities to the continuing growth of democracy.

Or we can look at Edgar Allan Poe, who was antidemocratic in many ways, and whose story "The Man of the Crowd" is all about how people in crowds become automatons. They lose their souls. It was a small step in Poe's mind, and in many people's minds at the time, from the crowd to the mob. There was a lot of fear of what was called *mobocracy.* Indeed, our whole representative system of government comes partly out of fear of the mob. If we just think about New York City, that fear was not always unjustified. Philip Hone, one of the city's earliest mayors, wrote in his diary in the early nineteenth century about how a Jacksonian mob rushed into City Hall and ransacked the place. Then there were the infamous draft riots during the Civil War, when mobs beat up blacks in the street. So urban crowds were associated with mobs.

Of course not everyone felt that way. Whitman certainly embraced the crowd, which he saw as a kind of natural phenomenon, like a waterfall. He waxed poetic about the crowd in much the same spirit that Wordsworth did about mountains or streams. But the mistrust and dislike of cities on the part of many American commentators continued until we get to Henry James, who returned to New York in the early twentieth century and was appalled at the immigrants streaming through Ellis Island. He couldn't understand how the Republic would hold, with all these swarthy types, these Italians and Jews and blacks, swarming in. What would happen to the native stock?—by which he meant not Native Americans, but people of English, Scottish, and German descent. So cities were places that attracted immigrants, which challenged the notion of a stable WASP leadership class.

Other opposition in the twentieth century came from city planners, who had a fastidious bent and could not understand how metropolises could keep getting denser and denser and expanding in more or less organic fashion. They ought to be planned and contained. Patrick Geddes in England and his disciple Louis Mumford in America championed the idea of garden cities. The idea was that we would break up the big cities and form new planned communities, such as would be attempted in Radford, Virginia, or in our very own Sunnyside, Queens, or more recently, by the Disney Corporation, in Celebration, Florida. These planners kept asserting that you can't just permit a city to grow—it's not rational. There was a feeling that New York City, especially, "growed like Topsy," that it was out of control, that it would have to come crashing down at some point. That was a position of both the Left and the

Right. I well remember, in the 1960s, how some of the hippies started going back to the land; rural communes were popular, and Paul Goodman stated that it would be a good thing if 20 percent of the American urban population were dispersed to the countryside.

More recently, in the 1980s, there were many dire predictions that New York City was at the point of extinction, or indeed that its death had already occurred, in some invisible, subcutaneous, but no less real way. So the feeling persisted among many progressives and conservatives that American cities were a doomed species, and rightly so. "The land" equated to innocence, the cities, to corruption. If we examine the iconography of the lyrics to "America the Beautiful," we see that there is no stanza about Coney Island. "Oh beautiful for spacious skies" and so forth employs imagery that has been exploited by politicians and advertisers, from the Marlboro Man, to Ronald Reagan's commandeering the image of the cowboy on the range, down to an even more recent president vacationing on his ranch in Crawford, Texas.

This open-space imagery carries with it the notion that Americans should be individualistic, self-sufficient, unreliant on government: "Don't fence me in." There is a steadfast refusal to acknowledge that America has been for a long while largely urban, a good chunk of its demographics clustered around spreading megalopolises. During his first presidential campaign, Bill Clinton never once uttered the word *cities* in any public speech. And I don't think he said it in the second campaign either, but I had stopped paying attention by that point. Imagine a politician issuing position papers on everything under the sun, and never saying the word *cities*. We do not have a national urban policy. We see, from what happened after the devastation of New Orleans, that there is an enormous vacuum at the federal level in terms of rebuilding cities. Who is going to do it, FEMA? Let's hope not.

We are a society that dare not speak the name: *city*.

Where do urbanists and environmentalists come in? I would say that most urbanists, most city-lovers, are basically environmentalists. They don't want to see the city they love polluted; they are happy when the rivers surrounding it are cleaned up, when they can breathe cleaner air. It may not be their ruling passion, but they are definitely proconservation and protection of the environment. I don't think we often find quite the same emphasis among environmentalists. Consider the fundraising literature that is put out by the Sierra Club or other conservationist groups. There are not many images of cities. There isn't much positive rhetoric about the need to join hands with our urbanist brothers and sisters. The Environmental Liberation Front, or ELF, does not spray-paint graffiti advocating infill in city neighborhoods. But infill is a sensible solution, energy-wise: build on those empty lots.

We have a tendency to think of density as congestion, and congestion as a bad thing. It's curious: some of the most densely populated blocks in the country, not just in New York City, occur on Park Avenue in the ritzy sections; but we don't think of them as blighted because they are large, broad-shouldered apartment buildings with doormen in the front. The whole crusade against tenements, from Jacob Riis onward, which was a noble war, was partly based on a misunderstanding: that the problem came from crowding too many people into too small a space. That wasn't what was wrong with tenements and ghettos, per se. What was wrong was the poverty and the inadequate food, medical care, hygiene, and job opportunities. So when the tenements were finally torn down and replaced by high-rise housing projects, surrounded by grassy areas and cut off from the street, following LeCorbusier's antiseptic anti-urban towers-in-the-park notion, it didn't necessarily improve things. People began to realize that it would have been better to rehabilitate the tenements and keep the familiar neighborhood streetscape—which is precisely what is happening now in places like the Bronx.

So we learned. There has been a change of thinking in the last fifty or sixty years, starting with the postwar period: for one thing, suddenly the organic, higgledy-piggledy arrangement of New York began to seem aesthetically pleasing: the one-thing-after-another of it, the tall building next to the short one, the liquor store next to the bank, the passing stream of faces in the street. An urbanite wisdom was developed by writers such as Jane Jacobs, in her book *Death and Life of Great American Cities*, or William H. Whyte, who did time-and-motion studies of how people actually move in the streets or utilize public spaces. Some of this street wisdom was just common sense: for instance, that it was preferable to maintain the traditional street wall, with retail on the ground floor, instead of building corporate glass towers set back from empty plazas; or that if you wanted people to use those forbidding plazas, it might be a good idea to put in some benches. These street-smart insights grew out of the conviction that what made cities worthwhile was their vitality. Jane Jacobs famously declared that a park that is unused is a dangerous park, and that a street with a tavern has a better chance of being safer because people will be going in and out of the bar and eyes will be watching the street.

Some of these points sounded counterintuitive at the time; others made perfect sense. The upshot was that a whole body of wisdom and practice accumulated about what sorts of designs worked to create a better, more alive, less dangerous, more pedestrian-oriented experience. In the meantime, New York City has gone through an amazing transformation, becoming safer in many ways that it had been for decades.

Photo by Betsy McCully

But another kind of fear has persisted, which I would call the fear of city-making. We used to know how to build cities, how to add convincingly onto the urban web so that the new parts seemed contiguous with the old. Part of that confidence had to do with scale: a block would be laid out into lots, and one building after another would be erected on it, until you had, say, twenty buildings of all different sizes cheek-by-jowl, and the whole would fit into the grid. By the way, I am a great believer in and defender of the Manhattan grid. I understand where the writer Tony Hiss is coming from, and I sympathize with him and other contemporary planners when they say that you need to recognize the topography of the land and try to fit the streets to its dips and curves, and not squeeze everything into orderly rectangles. But that wasn't done here, and the results were—magnificent. Manhattan has a great functioning grid. It's not a boring grid either, because Broadway cuts diagonally through the squares, creating triangles along the way, and the short blocks of the Upper East Side contrast with the long ones of the Upper West Side. . . . I'm not going to defend the grid any further. It was good enough for Mondrian, it's good enough for me. In any case, you had these streets, with one building after another, and out of that came a variegated cityscape that was very interesting. With contemporary construction, however, you have one enormous building with huge unbroken floors taking up a single block, which leaves an effect of monotonous, monolithic, isolated giants. It becomes harder to create an intriguingly collage-like streetscape.

So the question I'm raising is: What should the environmentalists' position be vis-à-vis proposed urban development? You can't just be against all development, because then cities decay. If you don't put any new capital into them they fall apart. I was in East Berlin before the Wall came down, and I saw what happened when no money was put into repairs or infrastructure: it looked like you could have knocked down some of those buildings just by leaning against them. There has to be reinvestment in cities eventually; there has to be the recognition that neighborhoods change, get gentrified or whatever, and that that isn't always a bad thing. I'm torn emotionally about this, because I love the spirit of old neighborhoods, particularly working-class neighborhoods; but I also realize that when an area has become deserted or semi-abandoned, when it's forlorn and not enough people are using it, it makes sense to start rethinking and altering it. But how? That's the question. And what should the environmentalist position be about these changes?

For instance, if we just look at the New York City waterfront, which is an area I have studied, we see that the emphasis of environmentalists has been to put a green ribbon around the entire Manhattan, Brooklyn, and Queens shoreline—basically, a continuous park. Now I like parks, and I think a park can be a very good idea at times, but it's not the only thing you can do at the water's edge. You don't have to put in this prophylactic greenbelt designed to make everything pastoral. People don't come to cities just to be bucolic. You can quicken the pulse at the water's edge: think of the big Ferris wheel in Vienna or London, or other activity magnets in Paris on the Seine. I think that we're afraid of building anything at the edge, and so the effort has been to keep all the land in cold storage by declaring it "park." By the way, all these parks retain a vaguely theoretical quality when they're opened: they still look like architectural drawings that have been pumped up. The trees, the berms, the circulation patterns, are all very tame and tepid, for the most part. In Hudson River Park, which is only slightly wider than a sidewalk, you have to edit out mentally the presence of the West Side Highway, or as it is called now, Route 9A. And for all the new waterfront "access," there are still barriers preventing people from actually putting an arm or leg in the drink. There is a kind of lingering paranoia about letting people get to the river's edge in this city

I would like to argue for a more balanced planning approach to the waterfront, which would include more water-based options and activities that are not just contemplative. Utilizing the rivers as a serious mode of transportation, with beefed-up ferry service and water taxis, would be all to the good. It would also help to plant activities or structures on the waterfront that will draw people like magnets, either because they're entertaining or practical. Finally, it might be a good idea to loosen up, compromise, even, and not be so self-righteously literal-minded in defense of nature.

There are these weird contretemps that occur in big cities among conservationists, urbanists, and preservationists. Stuyvesant Cove, now a lovely park on the East Side of Manhattan, once held a concrete batching plant. There was a kind of promontory or rock in the water, just offshore, made originally of concrete and riprap, and the Environmental Protection Agency wanted to have it removed because it wasn't naturally formed. But in the daily life of that neighborhood park it functioned exactly like a natural formation, and looked for all the world like the kind of rock a mermaid might sun herself on.

Another example: you may be aware of the fact that on the East Side below 59th Street, there is a straight-as-a-ruler bulkhead that gives the shoreline of midtown and lower Manhattan a very hard edge. I was talking to someone in an environmental defense agency about the possibility of putting in a softer, more undulating edge, which would look both more natural and more interesting. He said it would be a great idea, but you might have to do battle with the preservationists, because the bulkhead, which had been built in the late nineteenth century by the Department of Docks, had itself been landmarked, and so any tampering with it was technically illegal. Beyond that, you might have to do battle with the environmentalists because, in order to secure some vegetation at the edge, it would be necessary to put in a little bit of understructure. So, paradoxically, in order to create a softer, more natural-looking edge, you might have to do some (shudder) LANDFILL, which is a definite no-no. Forget the fact that there have been enormous amounts of landfill in

New York Harbor in the past; ever since the battle of Westway (i.e., the never-built Westside Highway), we can't even contemplate the idea of landfill, however minuscule, in the city.

One final question: Is it that environmentalists' city-loving side somehow doesn't get expressed? Environmentalists often live in cities, so why isn't urbanism part of their conscious program? Why isn't there more common cause between urbanists and naturalists about how to keep what's vital in city life still vital, and how to fight for development that would provide jobs and be at the proper scale, and resist development that would be monolithic and soul-destroying?

Ever since I wrote my book *Waterfront*, I have been asked periodically by the *New York Times'* Op-Ed page editors to sound off on the coming of an IKEA store to Red Hook, or the Brooklyn Bridge Park, or the Atlantic Yards project. My response is to put my head in my hands until the offer goes away—because these projects are so complex, and the specifications keep changing, and sometimes the neighborhood groups fighting them sound as irrational to me as the developers seem greedy. I can't situate myself on a high enough plane to be able to judge clearly; try as I might, I can't get the broad picture. I can only keep enunciating what to me are some city-loving standards and criteria. But, for instance, I don't know when God declared that there should be no skyscrapers in Brooklyn. The skyscraper is a modern form; why can't you build a skyscraper in Brooklyn? All this is very confusing, and I'm proud to say that I didn't want to bring solutions, but only to muddy the waters. I hope I have done so.

Can You Eat in Soup?

Nine Million Ways to Look at a Raccoon—and an Apple

David Rosane

The universe is made of stories, not of atoms.

—MURIEL RUKEYSER

My friend John Waldman has asked me the following question: "Does nature matter to New Yorkers?"

I asked him, "Is the Pope Catholic?"

My first premise: the world as we know it is going to hell in a handbasket. Our entire planet is in a state of overshoot. In one year we consume what the earth needs a year and three months to produce. To bring the world population to North American consumer levels would require four additional planets.

My second premise, a quote from Canadian zoologist David Suzuki: "We are the environment, there is no distinction . . . just a big blob of water with enough organic thickener added so we don't dribble away on the floor."

It follows that if we are to "protect" or "preserve" or "conserve" or "save" the environment we must also "save" ourselves—and a lot of water. By way of logic, saving nature becomes a humanitarian project.

Ah, to save the world . . .

Not only is this a radically Western idea, it is preposterous. We are the fruit of fifteen billion years of cosmic time; of slow, gradual increments in complexity; of four billion years of painstaking biological change. As a species we are way less than a million years old. Dust in the wind. Who are we to have the power to "save" the environment—or even ourselves?

Do New Yorkers even matter to nature?

The universe will go on, with or without us. As a species we will one day "be no longer"; we may morph, evolve into something else, or go extinct. In any event, the "Anthropocene" will have come and gone. So will have industrial civilization. What's left is our past, the present, and as much of the future as the future has in store for us.

Protect nature? Why bother?

Thank heaven for New Yorkers—they've actually helped me come up with a few possible answers.

I'll start with a recent college student of mine, Joshua, an orthodox Jew. Joshua was part of a typical multiethnic college classroom from Queens, including Latinos, African Americans, Asians. One day we were on a field trip to Inwood Park, on the northern tip of *Manahatta*, the "island of many hills," as part of an urban nature course. We had hiked deep into the tall woods of "the Clove"—a valley full of 120-foot-tall tulip trees and giant slabs of half-million-year-old schist, plus one Indian cave where the Lenne Lenape used to hang out. The park was blanketed with giant piles of fresh snow and I was showing the kids how to hand-feed wild birds—chickadees and titmice and woodpeckers. The class was bursting with excitement. The birds, singing their heads off. Joshua happened to be staring at me—he looked a tad perplexed. A chickadee landed on his hand, took a peanut.

"What's up?" I asked.

"Oh, I'm just wondering, professor, are you an 'environmentalist'?"

I sensed by his tone that Joshua had been taught to be wary—very wary—of tree-huggers. Weirdoes from the Left. At the same time, I could tell he liked me, that he loved the course—he could touch and feel and identify plants and animals for the first time. Everything I showed him inspired sheer gratitude. Every color, every taste or scent or nature-induced introspection was a revelation, and cause for celebration.

Hence his dilemma. I tried reassuring him: "Joshua, I am not an *environmentalist* in the usual sense. That's because I don't believe we can 'save' the environment. My vocation is to reconnect you guys with nature, as best I can, *while* I can."

Joshua gave me the befuddled frown (the one with the open mouth and the nod). I continued, "Ok, take the fact that you and I are hand-feeding wild chickadees. We're totally entranced by these little guys, they weigh 10 grams and they have as rich an evolutionary history as ours, and we're totally digging it. I live to share these interactions, this kind of 'epiphany.' That might make me more of a humanist than an environmentalist. A lot of *enviros* think humanity is a disease, a virus, something we should get rid of. I beg to differ. I love your smile as much as the chickadee in the palm of your hand."

Joshua beamed back at me.

"Plus, you and the bird have just contributed to each other's welfare. You get to feed a chickadee while he puts a smile on your face and makes you go 'wow.' Think about it."

"Wow . . ."

I beamed back. An idea hit me.

"Joshua, the pleasure is all mine. Our little chat has just given me a brain climax: if our societies could shift their focus to the actual *relationship* with nature, and value the interaction itself, its potential for reciprocity, we might end up protecting both parties, indirectly. We'd be doing two birds with one stone, as it were."

I thought some more, fed a titmouse.

"I see one major hurdle. Our minds aren't used to thinking systemically. We see the world as a series of isolated individuals, objects, stereotypes, we think of 'one'—or the 'other,' 'Man' versus the 'environment.' We're a society that tries to save trees, or human communities for that matter, by putting gates around them. We're completely divorced from reality. How could we possibly save a relationship?"

Joshua insisted that relationships, *all* relationships, required commitment, chutzpah! *His* people could pull it off.

The rest of the class cracked up, then chimed in, arguing youthfully that we could *all* "save the world" if only we just reconnected and grew *with* biological nature, instead of trying to grow *out* of nature, away from it, as we have been doing in the West for millennia, destroying the earth as a result.

I acquiesced, save for one caveat.

"Guys, all of us, the environment, the whole system, is governed by turbulences, feedback loops, nonlinear dynamics, complexity. Nature cannot be 'managed,' nor told what to do, by some linear-thinking primate, no more than 'it' can be 'saved.' Life equals change—and it will change, on its own terms."

"Now I know why New Yorkers are always so upset with the weather, professor."

"Why's that, Joshua?"

"It's the one thing they can't control."

Claude Levi-Strauss once tinkered with the idea that cultural tolerance is not contemplative. It is a verb, an action. A dialogue. Not only was

Joshua enthused by the coherence of our conversation, it also fit cozily with his world view. Even he could be an "environmentalist"! For a moment, our respective ideas had come together to forge a new and mutually acceptable proposition; we were like algae and fungi, connecting to make lichen. In nature, different parties are always merging and generating that which they would have been incapable of achieving on their own.

How about we apply the same strategy to our current "situation" with the biosphere—resume bilateral talks? Say we exercise some of the same built-in propensity for creative intercourse. Some natural diplomacy.

First, before we innovate, there's some *listening* we have to do, some redefining. Ecology is the study of all the interactions, from the cellular level to the global, that shape and characterize an organism. It follows that nurturing our *relationship* with nature (i.e., our ecology) means interpreting all of our planetary ties, means first knowing the sum of our interactions with the world—and their consequences. Keyword: awareness. Method: education. Goal: ecological literacy, a fluency in the

Photo by Sarah Muehlbauer

language of life. Enter Mike Feller, Chief Naturalist for the Parks Department, mentor and friend, the guy who first introduced me to the wondrous art of nature shpiel and the planes of ecological consciousness. Mike's keyword: "Process!"

A few years back, one April, we were walking with his wife and kids in Alley Pond Park, in Queens. A slice of green, wedged between two highways, courtesy of the automobile and the late Robert Moses. Mike pointed to his two daughters who were hurrying ahead of us, up the slope of the moraine, trying to beat us to the spring peepers peeping away in the park's glacial kettle ponds ahead, to the bellwort and the bloodroot, to everything. He delighted in their drive, their resolve, their eagerness—their ingrown "pursuit of happiness."

He smiled, speaking softly through his beard: "You know, you hear all these lofty reasons nowadays for protecting the environment, 'climate change,' 'species loss,' economic reasons, long-term profit motives . . . the only one I need, but never hear invoked, is the 'life experience.' "

"Say again?"

"Well, take all this hype about biodiversity; we now know that species are only part of the picture. What's also at stake are billions of *processes*, the interactions among organisms, events like photosynthesis, respiration, decomposition, the carbon cycle, the water cycle, human languages even. That's what needs 'protecting.' It's the processes that keep us ticking, not the parts themselves."

"Relationships, in other words?"

"Yeah, and the experience thereof . . ."

His eyes twinkled. I scratched my head. He continued:

"When you feel your own breathing, when you taste what you're eating—that's life experience right there, and it carries intrinsic value. Everyone's experience has value, for that matter, from worms feeling their way through soil, to a mole feeling its way to the worm."

I argued how difficult it might be for a materialistic culture to uphold the existence value of something as immaterial as, say, the experience of bug-ness.

Mike disagreed: "What about our flaunting of liberty, God, or consumer satisfaction?—these are equally intangible properties."

Mike explained his philosophy, a partnership ethic with nature that transcended both the anthropocentric and the *eco*centric (and definitely the egocentric). "Simple stuff, like planting tomatoes or growing spicebush at home, or raising bees for honey, or restoring a saltmarsh in

Brooklyn and involving the local neighborhood. You know, peacemaking, the sort of deal where everybody wins. Love for all, exploitation of none. Nature as kin. An eco-repeat of Gandhi's 'To live simply so that others may simply live.' "

"A culture of biology?" I ventured.

"Aren't they supposed to be the same thing?" he asked.

Mike is my Socrates. He's also right about his *processes*. I checked the literature. Life on earth is autopoietic, or self-creating; it is both the producer and the product of unending chemical and physical "conversations" that connect cells within individuals, individuals within ecosystems, and ecosystems within the biosphere, thanks to which species *do* come into existence, *relate*, then fade away, only to be replaced by others. The resulting fabric is a planet-wide matrix of intertwined diversity, networks of beings, resilient communities—all of them in translation, flux, animated by an infinite and relentless and equally meshed undercurrent of matter and energy on the move. Metabolism, from the Greek *metabole*—change, from dinos to who knows . . .

We passed some beech trees, spotted a Great Horned Owl, found a spring peeper, knelt to touch some early geranium blossoms. We imagined their hidden connections.

"I assume you think the Creation does exist, just that it's an ongoing process?" I queried.

"How about a *play* in progress. Think of Earth as a giant Ferris wheel; matter and energy, revolving in cycles, taking on new passengers with every revolution."

Mike is from downtown Brooklyn—used to spend summers at Coney Island as a kid. He knows his ecology, his plants and his birds and his entomology, as well as any modern-day Thoreau or Leopold or naturalist from Vermont or Ohio. He continued to talk of his daughters' innate capacity to explore, to learn, to giggle, to interact with a meadow full of wildflowers and butterflies. To roll in the grass or in the sand and sneeze their heads off and climb and fall out of trees. To bathe in pollen and fragrance and bask in their own ability (and liberty) to respond, to integrate and elaborate on the world around them.

Everybody's capacity, for that matter. Everyone's right, too.

We *are* the universe. There is no distinction. A blob of water (brought to earth by inbound comets) with enough organic thickener added so that we get to smile at a bird or smell a rose or shower in the rain or run away from a man-eating tiger and then stop and stare at the stars. And reflect on them.

It follows that to interact with our environment, to get to know it and "engage the universe," as Albert Camus commanded us to do, is to lay claim to the kind of symbioses that honed us and built us over cosmic time and that today define us as human beings.

(Ultimately, it is to retrieve our humanity.)

It follows that to "be the environment," as Suzuki reminds us, is to reclaim ownership of our most basic and most fundamental *human right*—the right to exist within a healthy web of relationships, to live within a *functional* ecology.

It follows that we are theoretically entitled to a hale and hearty world, to breathe clean air and drink crystalline water and eat real food and to flourish, unimpeded by environmental destruction or industrial toxicity.

It follows that we can and must defend and nurture our "ecological citizenship" as we would defend our other most basic rights of citizenship.

It follows that to be stripped of this right to be *of* the environment is to be denied our first right to existence as living individuals.

It follows that to have one's environment destroyed or impaired by another human or human entity, to have one's *relationship* to the environment destroyed or impaired by a second or third party—to be handed poisonous food or toxic air or dirty water—is to be the victim of a heinous crime. More blatantly, of an abuse of our most elemental human rights.

Does nature matter to New Yorkers? It matters to protect our first and oldest human right to be a part of it.

How to belong, fellow Gothamites? The goal for us scientists, naturalists, educators, citizens, is to understand and share and emulate the sum and sequence of these "stories" that link all of biological diversity—the "energy flows," the "nutrient cycles," the grammar and vocabulary of ecological design. To learn how stuff grows, basically. Not just corn or rice—entire ecosystems.

New Yorkers, to your books and to your hoes!

Ecologist Paul Mankiewicz of the Gaia Institute says there's enough water generated by this city to sow a temperate rainforest throughout the urban tapestry. Let's do it. Let's throw in a couple million more bicycles and a better, non-screeching public transportation system, too, for the sake of sanity, our cochlea, our lungs.

What else? Monopoly man, cough up some green. Invest in community gardens, greenroofs, local agriculture, giant piles of mulch. Don't forget anaerobic digestion, either (waste into fuel—very cool). Recycle all materials within a service and flow "economy of scope."

Green economies are inherently creative; they *will* create jobs.

And by the way, since life is change, then even we can change. We have before. Let's believe in evolution.

Let me share a few things about my own natural history. I was born in Guyana, grew up hop-scotching from Canada to France to Spain and back. My immediate family is from New England; my father worked (and our family lived) abroad for a North American transnational corporation. I've been a lifelong expat. A "corporate brat." A lifelong naturalist, too.

These past four years living in New York have been my first time living in the States, my "home country." My first time, too, as a student and teacher of the urban environment, studying and teaching and writing about nature and people in the city.

Immediately prior to entering the Apple, I toiled in the rain and cloud forests of Venezuela and Peru as a research assistant in ethnobiology for Cornell University. I'm used to things like Harpy eagles the size of motorcycles that fly around ripping monkeys out of trees, jaguars that "kill with one leap," butterflies the size of dinner plates, and spiders the size of your face. For seven years I studied these birds, insects, mammals, and plants—and their relationship to hunter-gatherers. Was such-and-such a fruit used for food? Or for medicine? Was this or that animal an important reference point in a people's culture?

My job *hasn't* changed (that much).

My first spring in Manhattan I spent bird-watching in Central Park, mostly in the north end area, up the Great Hill, around the Pool, down through the Loch and into the Ravine, what Olmstead designed as the Adirondacks of this great public space. Most of it is woodland or meadow; all of it is "planned," none of it "wild." Surreally manicured within the concrete, overraked, overfertilized—an economic and ecological sink. A puddle of green, assimilated by the grid.

In early morning (5:30-ish), male prostitutes (and their clients) and birders are the people you usually first bump into. Then come the dog owners, the joggers, the families. The cops and the gardeners. The occasional turkey, the rare coyote.

One particular morning I will carry to my grave. I had been in the park three hours already, since the crack of dawn, and a warm breeze had New Yorkers giddy-faced and flushed. Bugs were buzzing and birds everywhere were bubbling over with song. I was ecstatic, on a "warbler" high. Spring migration was peaking, big time. The vegetation was teeming with high-strung, feathered dinosaurs, migrating north from South America, stopping here to feast and refuel, many of them on freshly hatched inchworms. My eyeballs were drunk with the likes of Rose-breasted Grosbeaks and Scarlet Tanagers—usual fare for Central Park on a good day in May. There were Indigo Buntings, too; a perfect day, really. The sun was shining, the pin oaks and tulip trees glowed lime green, the sky was blue. But what had me really happy was my warbler count: more than twenty species, including Blackburnian, Cerulean, Golden-winged, Blue-winged, Chestnut-sided—Miro paintings, with wings.

Too bad I had to leave the park for a meeting in midtown.

I was exiting on a path lined with benches. To my left, just ahead, two women were seated, each with two young children at their sides. The kids were bouncing with enthusiasm. The two working-class moms, one Hispanic and one Asian, were waving to me, smiling profusely, gesticulating. They wanted me to slow down and to look upwards, up behind my right shoulder. I did. Whoa! Male raccoon. Old, raggedy, sick-looking, male raccoon. Hunched up in the fork of a tree.

My first Central Park raccoon. I love raccoons.

I inched my way over to the bench, said hello to the women, introduced myself, then sat down beside them. Smiles all around. They explained what was going on in broken English. They had been watching the animal for minutes, waiting for the old geezer to climb down. At the base of the tree was a public waste can and he had been sizing it up with his beady, glazed-over eyes.

We waited. So did the raccoon. At last he climbed down, head first. The four kids went bananas. The raccoon took his time. He arrived at the base of the garbage bin, climbed up its side, laboriously, and then disappeared inside. We heard rummaging. The kids held their breath. He emerged, holding a paper bag from Burger King. Junk food (how surprising). Screams of excitement all around. The Latino lady clapped her hands; the Asian woman said something to her kids.

The old raccoon climbed back out of the garbage can, sat at the base of the tree and, facing us, proceeded to adroitly unpack and throw back half of a leftover Whopper.

Bursts of laughter, all around.

Rewind. Here we have a very white naturalist of European descent (me), two beautiful young mothers from foreign lands and their kids, all awash on the shores of springtime glee, brought together by one very native and very adaptable American animal salvaging the worst food known to man from a waste can cloaked in derived petrochemicals in the most contrived slice of nature this side of the Atlantic.

Exit Miro, enter Hieronymus Bosch.

And welcome, by the way, to *the* nature of New York.

Icing on the cake? As we laughed with the raccoon, a heavily powdered lady in high heels and slacks and a small designer purse hurried by. Country of origin: Upper East Side. She jumped at the sight of the raccoon, then without stopping, spouted out some confused advice about "not letting the kids get too close to the animal, they might get rabies or West Nile or the bird flu even."

The Latino mom, to my right, sighed and then shouted after the woman, "The kids are fine!" She looked at me and shrugged in exasperation. "*Cogno!* These white people, they spend waaaaay too much time in libraries!"

The raccoon finished up and climbed back into his tree. He started to clean his coat, like a cat. I turned to the Asian woman, who was looking at the raccoon. She noticed me, smiled, and then timidly inquired, pointing at the animal, "Can you eat in soup?"

Here in the city I do exactly what I used to do in South America—I study plants and animals and their connection (visible or not) to people. Has my "environment" changed? Sure, I switched jungles. Here's the fundamental difference between the two: the rainforest is a "wilderness" of millions of species of plants and animals, including top-of-the-food-chain predators, involving a gazillion processes, including humans—thinking, talking, laughing, storytelling humans with hyper-sophisticated cosmologies, living in ecologically sound societies. Places like the Amazon create life. They sing. Madhouses like New York absorb life, swallow it whole. City as vacuum cleaner. It rumbles, it heaves, it coughs, it's an engine of the global industrial economy. It's *supposed* to be noisy and cantankerous and brash and irreverent—it's a giant machine. John Steinbeck wrote: "New York is an ugly city, a dirty city; its weather is a scandal and its politics are made to scare children."

Not only is New York the most hurried and competitive place on earth, it always has been. It was the first city in history to be founded by a corporate power system. It wasn't "settled," it was "opened up," like an office, by a private trading outfit, the Dutch West Indies Company. The Dutch didn't accept people of different creed and nationality to

New Amsterdam because they were nice but because it was good for business. New York City—the Capital in Capitalism—whose mantra continues to be, "A competitive environment is a healthy environment." Nature, red in tooth and claw. The law of the jungle?

I'm a naturalist (and a humanitarian) who revels in collaborative mutualism, who understands that "survival of the fittest" is only half the picture. The other half? Pollination, seed-dispersal, symbiogenesis—the merging of two separate organisms to form a single new organism. Lichen! Reciprocal altruism is what keeps my doctor away (not apples).

So New York City took getting used to. It still does—after four years I still feel new to the place, a brittle-skinned tenderfoot.

Wait a minute. New York is all about being "new," so technically I might still qualify as a *typical* New Yorker. The city's last census found that close to 40 percent of us New Yorkers were born abroad. Newcomers. *Imigrantes*. The place has never been so ethnically diverse, nor socio-economically complex, as today. We're talking people from every tribe, ethos, nation, class on the planet. Remember this factoid: over 60 ethnic groups, writing in 42 languages, publish some 300 non-English-language magazines and newspapers in New York City. There are more languages spoken in Queens than there are nations represented at the UN. The world is the city's tributary (for better or for worse).

Does nature matter to us?

Try to imagine what would happen if we asked John's question to everyone out there. In as many different languages as are spoken here. In as many different world views. How many of the sixteen million people living and working in metropolitan New York actually have a word for "nature"? Are "people" included in "nature"? What do Haitians consider their beheaded voodoo chickens to be—the ones that end up floating down the Bronx River? And the watermelons placed at the bases of trees for ritual, throughout most of our city's parks? The turtles released into our parks' ponds and lakes and streams by New York's Buddhists? What do the Guyanese Hindi think? On a beach in Queens, adjoining Crossbay Bridge, they pray and leave coconut and hibiscus and banana offerings in the water and on the sand. They wade out up to their hips in the cool waters of Jamaica Bay, come rain or shine, smack in front of the thundering take-offs at JFK. What is "nature" to them?

A majority of cultures do not distinguish between "animal" and "meat," "environment" and "home." Neither are objectified—both are subject. Full of spirit. Full of breath. Full of *psyche*. We are the environment. There is no distinction possible. I'm betting my last rupees that your average *typical* New Yorker considers himself to be part of nature too. So, do New Yorkers matter to New Yorkers?

Back to my main point: the environment, as we like to call it, has an impact on who we are and what we do, and vice-versa. We come here today, to this city, and nine million of us drink the same water. Little do we know that we thereby qualify as an "ecological" community, bounded not only by the same exhaust fumes, ozone pollution, noise pollution, light pollution, endocrine disruptors, carcinogenic particulates, and de-vitaminized, de-mineralized, industrial food, but by the same commons of mountainous, hardwood-covered hills—our watershed. The Dom Perignon of public waters, straight out of the Catskills, today served with a peppering of mercury, courtesy of our midwestern coal-burning power plants.

Here, at this meeting of river, highland, estuary, island, and ocean, we metamorphose. Big Apple as chrysalis. As feedback loop: when our surroundings change, we change; in turn, we modify our surroundings.

Can our city change the planet?

I have a huge confession to make.

When I first came to Manhattan I originally expected nothing more than the cliché setting—a few rats and a bundle of pigeons in an otherwise hypergentrified world. Disneyland, without the parking. The idea of "nature in New York" sounded at best an oxymoron.

But harp seals in the harbor? The highest density of breeding peregrine falcons in the world ? Plus one million mites (and a new species of centipede) per square foot of Central Park soil? Come on . . .

Perhaps I was assuming that life cannot live in such sterile conditions. Well, what about us? And the rats, plus the ones with wings? Nature is adaptable and life all-pervasive. Maybe that's why so much of it actually *does* adapt to New York. The idea dawned on me that I could simply reverse the question. "Why is there nature in New York?" became "How come New York City was built on top of so much nature?"

Life can survive as bacteria miles beneath the surface of the earth, or as tiny insects called springtails that live on the ice of Antarctica, or as bedbugs in our sheets in New York. Animals can migrate thousands of miles, switch continents, staple entire landmasses and oceans together. Every year tens of thousands of migrating shad swim past Battery Park. They swim to spawn in the Hudson. In New York you can also see monarch butterflies flying past your window, in the fall, en route to Mexico.

In fact, this city is home to three thousand documented plant species (including three and a half million trees) and an estimated ten thousand animal species within the contours of the five-borough Apple. Some species are native, some foreign; some elevated to star status (Pale Male), others threatening—and therefore reviled, like poison ivy, the Asian longhorned beetle, garlic mustard.

Most of them however, are simply unknown, or ignored, by nine million *Homo sapiens.*

Illiteracy? Blindness? How about a combination of the two: Annie Dillard says that "seeing is an act of verbalization," that "we only see what we can call out by name."

I'm a naturalist, published researcher, and environmental educator, but I didn't realize that there was such vibrancy of life in NYC. I *can* call stuff out by name. I *am* supposed to know these things, *was* supposed to know these things. I didn't.

Why was I amazed?

My high-school students have helped me craft some nifty answers. On a field trip to Jamaica Bay, one student, while observing a Great Egret, asked, "Who put that bird there?" When it came time to engulf our picnic lunch, our tuna sandwiches, one kid asserted that "food can *no way* be defined as nature."

Generally speaking, children who have just immigrated, especially those who come from rural areas in their country of origin, are not astonished to encounter "wild" animals in the city. Those born here, in New York, tend to carry all the predictable biases, the preconceived notions; a majority compare squirrels to thieves, describe nature as lethal. Again, violence and its immediate corollary, our culture of fear. Nature fear. Fear of everything. Somatized fear. I've had students allergic to grass, literally; others who, on their first field trip, feared for their hinders—one girl affirming that she had never "sat her sweet ass on a rock before." More proof, I guess, that with urban environmental education you've got to start at rock bottom.

And what if fear had everything to do with it? Nature as the enemy? City as safe haven? Raccoon-fearing white people in libraries!

Marcel Proust said, "The real voyage of discovery lies not in seeking new places but in seeing with new eyes." My entry into New York City has been a journey beyond simple opposites, to the land of nuance and ecoliteracy. I have learned that from the standpoint of Mike Feller's life *processes*, cities (and their raccoons) are neither "natural" nor "artificial";

they lie at one relatively sterile end of the global life continuum, with fully functional "wilderness" and species richness (in places like the Amazon) lying at the other. Somewhere along that gradient, somewhere in the middle, are our urban refuges, parks, sanctuaries. Gated communities, chunks of nature embedded in a metropolitan matrix. *Bon appétit!*

The Dark Side; or, My Time Spent in the Nature That People Would Rather Not Think About

Robert Sullivan

If nature were a political candidate, and if newspapers and television networks took surveys of the public's opinion of nature, then nature would, at this moment in the twenty-first century, most likely have, to use the pollsters' phrase, high positives. For the majority of Americans who live in suburbs or the various rural and semi-rural permutations of the suburbs (exurbs and ruburbs, as they are known), nature looks good. It's the place where we like to vacation. It's the view from our calendar. The place we get away to—literally, or in video or in books and magazine of all outdoorsy kinds—after spending most of our time commuting to work on the highway or driving from the mall-esque school to the mall.[1]

Nature is not generally associated with the city, in other words, and when it shows up there, people can be overly complimentary about what is always referred to as its "persistence," or we treat it as if it's on its last legs. In the popular view, nature makes cameo appearances in the city—in the park, which we view briefly while on a stroll at lunchtime, or on Sunday with a barbeque, or in the view across the street at the squirrels and the pigeons nagging around the free lunch thrown from the bench. In the city, we feel that nature must work overtime to survive, or

[1] In the landscape of the road, which is the predominant landscape in America, seeping now from the edges of the highways into the center of cities everywhere, like an invasive species of grass or vine, the school also looks a lot like the just-as-ubiquitous landform, the prison.

at least work harder than usual, just to break green through a crack in the sidewalk (an example of reverse anthropomorphization). In the city, nature seems to be confined and out of place. When nature does turn up, it's often not card-carrying nature; it's the bad kind. I'm talking about bugs, pests, vermin. This is the nature that is not considered natural. In the city, nature tends to show its dark side.

That's what people seem to think, anyway, and I know, for I have been to the dark side of nature. I have spent a year in the city closely studying rats and all that is associated with them. I have studied rats in alleys, in basements, in sewers, and even in fancy restaurants and in nice parks where people don't seem to think there are rats, and, thus, feel free to toss their lunch out for the benefit of what they are ultimately less likely to feed, the pigeons. I went to the dark side, in other words, and, to be overly dramatic about it, when I came out I saw the light. Or I saw *a* light, or, anyway—but I am getting ahead of myself. What I mean is I saw the nature that everyone is not looking at, the nature that Ansel Adams avoided, that people don't use as screensavers or put on the side of their coffee mugs.

Before I explain my journey to the so-called dark side, I suppose I should say a few words about my own natural habitat, so that you might judge me by the nature in which I was raised. Not that it matters, but I come from the East, was born on the island of Manhattan, reared on Long Island, a half a block from the edge of the borough of Queens, also on an island, in this case Long Island. When I was eleven or twelve years old, I worked hard every Tuesday night for a year to acquire the Boy Scout Wilderness Survival merit badge—by trying to start a fire with a piece of old flannel pajamas and some London planetree twigs in a backyard in an Archie Bunkeresque neighborhood in Queens Village. I wanted to camp. I wanted to camp very badly. I had read magazines about camping, chiefly *Boys' Life*, and I had studied the Boy Scout manual for a long time, reading the camping directions (directions being what they literally were) with a mixture of bubbly excitement and nerve-wracking anxious fright. *Camping! In the woods! What do they mean by woods, anyway?* And like my father before me who did not camp but who grew up in New York when there were still farms and ponds on the fringes of the outer boroughs, I grew up believing that one day I would camp, though when my Boy Scout troop actually went on the trip it was less like what I now consider camping and more like camping out overnight for concert tickets, and all I remember is eating the emergency rations (candy bars) somewhere in a nonrural section of New Jersey late at night in the pouring rain.

I also grew up never really leaving New York and the cities of the Atlantic coast, but at the same time I loved to explore. My father was the prototypical urban explorer, at a time when his species had yet to be identified as such; my childhood was spent poking around in old about-to-be-destroyed buildings in Lower Manhattan, pulling over to check out old farms and empty fields in Brooklyn and in Queens. Thus, as a teenager, in the years when I was finally freed from the need to camp, I took my pent-up camping energy and applied it to the fields at hand—in this case, brownfields, as in disparaged industrial swamps, as in all the wastelands that characterize the edge of many cities. Cities tend to be set up near water or waterways, and the city's dumps and airports and industrial zones tend to be built on that land that has long been least valued, the swamps; and as industry left the cities, during my childhood in the 1960s and '70s, so a new kind of wilderness was born, a postindustrial wilderness that, amazingly, since the '80s and'90s has been paved over by superstores and so-called box stores and strip malls and giant entertainment complexes.[2]

In particular, I loved to hike and explore in the area that the people of the New York metropolitan area and football fans all over America know as the New Jersey Meadowlands. Famous as one in a list of possible burial sites of the infamous union boss Jimmy Hoffa, the Meadowlands were, for me, a kind of reverse natural commute. While people were racing to leave the city on weekends and get into the country, I was in the city, or just on the semi-abandoned edge of it, canoeing through polluted nature—a nature so disparaged, in fact, that most people did not consider it nature at all. I explored it for fun on weekends, and while working newspaper reporting jobs in the area, and even with my girlfriend, until we were married and moved away to her home state, Oregon.

When I arrived in Oregon, the scene was a shock to my noncamping aesthetic. In New York, I had dressed for work (coat and tie) or to play touch football in Central Park (jeans and college sweatshirt). In Oregon—or more generally, in the Pacific Northwest—at the beginning of the '90s, the trend in fashion was outdoor technology-related—i.e., the general public wore jackets made of fleece and so-called performance-style water- and wind-proof shells. People were dressed, in other words,

[2] As softwoods replace hardwoods in the forest's succession of trees, so low-paying jobs at the discount stores replace relatively high-paying ones at the midcentury factories and mills.

to camp. (This trend would move east a few years later.) I was in Boy Scout heaven. My wife and I went hiking frequently, and I made a couple of hiking friends and we all hiked a lot and talked about it even more.

Eventually, I even went camping. I went camping a lot, making up for lost time in Queens. I experienced nature, at long last, recreationally. I experienced nature the way a lot of people experienced nature then and experience it nowadays—by driving to a store, buying a lot of really cool camping stuff, driving into the mountains, hiking all the cool camping stuff into the woods, and returning, wearing the same clothes to work, and even to dinner and restaurants. By wearing these things, by parking at the coffee shop with our kayaks on our cars That these clothes were made of petroleum products, that the cars attached to kayaks contributed to the pollution that obscured the very stars beneath which these same nature-loving people sought to relax and sleep and burn wood (creating more star-smearing carbon), was not a big deal at all.

I loved camping, and I loved the classical wilderness experience, as opposed to my industrialized nature version, my reverse commute. I loved hiking for miles into, say, the glacier-covered Cascade Mountains, into an old-growth forest, and then coming on a waterfall and seeing no one except the occasional hiker wearing the same kind of clothes and carrying the same cool gear I was carrying. I really loved loving camping. I felt like a Romantic poet. Also, I loved my gear.

Wilderness, as a human conception, has to do with memory, to some extent, as well as with ideals and aesthetic values. Having started out with an idea of what wilderness was, I began to miss my old East Coast postindustrial wilderness; I began to ponder applying the classical camping skills I had honed in the West to the reverse-commute wilderness experience of the East. Pretty soon, after some finagling of frequent flier miles, I was flying east, canoeing through toxic waste sites—while wearing really good gear. If only in the wearing of the really good gear, my western and eastern experiences had been reconciled. I was paddling with a friend of mine from high school, a distant relative of Meriwether Lewis, believe it or not—to my mind, such as it is, my friend's presence gave our expeditions a kind of naturo-literary credential—not that *naturo-literary* is an actual term.

And it *was* an adventure, a legitimate exploration of what was forgotten due to its being close at hand, an exploration of a land that because so many people had seen it so often had become unseen. In creeks that felt a thousand miles from anywhere but were just a few miles from the Empire State Building, we met people swimming in water that I would not want splashed on me. We saw toilet fixtures on an island made of invasive plants. We got lost and had to be rescued by a security guard who put our canoe in his pickup truck. We came upon strange-looking and, because of my mindset, radioactive-seeming ducks (they were originally from South America, as it turned out).

In adventuring in the Meadowlands, I attempted to apply the reverse commute to my perception of so-called nonnatural things, those views that would not show up on calendars. At one point, for example, I attempted to follow a toxic stream up to it source, on a garbage hill, in a swamp where the water was what is formally known as *leachate* but is also known colloquially as *garbage juice.* I used the devices of classical nature writing to explore a nonclassical nature, a.k.a. nature's dark side. An example:

> One afternoon, I drove back through a field of abandoned cars and walked along the edge of a garbage hill, a forty-foot drumlin of compacted trash that owed its topography to the waste of the city of Newark. On the side of the scraggly grass-covered hill, little black patches were scars from a recent fire. There had been rain the night before, so it wasn't long before I found a little leachate seep, a black ooze trickling down the slope of the hill, an espresso of refuse. In a few hours, this stream would find its way down into the already spoiled groundwater of the Meadowlands; it would mingle with toxic streams, and perhaps dilute whatever rare drop of water in the region might somehow be without the trace of humankind. But in this moment, here at its birth, at a stream's source in the modern meadows, this little seep was pure pollution, a pristine stew of oil and grease, of cyanide and arsenic, of cadmium, chromium, copper lead, nickel, silver, mercury, and zinc. I touched this fluid—my fingertip was a bluish caramel color—and it was warm and fresh. A few yards away, where the stream collected into a benzene-scented pool, a mallard swam alone.

You might think that what I was doing was just a stunt, and on the one hand it was—on the fun hand, to be completely honest. On the other hand, the expeditions made it clear to me that nature was there, even though it wasn't advertised as such; canoe through the streams surrounding a garbage dump and you will experience wilderness, all right, just not the wilderness that you see on people's screensavers.[3] The Meadowlands

[3] I am pleased to note that things have changed somewhat in the Meadowlands. Where we once semi-risked our lives to put in our canoes, there is now a park with a section for

was (and in a lot of ways still is) wild, and there is an argument to be made that it is even more wild than the typical crowded national park, bears eating garbage behind the restaurants, hoards of hikers running you down, the buzz of all-terrain vehicles filling the air. Additionally, the exercise of the reverse commute has benefits of its own. Just as you can see the weekend vacationers pulling out their hair in returning Sunday-night highway traffic, so you can see what's a little nutty about setting up a tent in a field on a mountain, these days bringing your solar-powered cell phone with you. Likewise, you gain greater insight into the wilderness that surrounds you, the wilderness that you had not necessarily noticed before, when you are not in a dark forest of trees.

I continued to live in the Pacific Northwest after I went on my expeditions into the Meadowlands, and soon enough, I returned to classical camping, hiking to glaciers, to mountain lakes, to the source of crystal-clear, leachate-free streams. It is, of course, true that acid rain touches everything, that global warming means that man has tainted all and there is nothing that is "pure," though to think that things were once pure again places man out of the equation of nature, which, as I am attempting to show, is, I believe, misguided.

One fall, I ended up camping on and off for almost two months in the Olympic Rain Forest, on and around an Indian reservation, the home of the Makah, a Native American tribe that was in the news for a time for having hunted a whale. Though the tribe waited until the whale came off the endangered species list, though the tribe only used the whale meat for ceremonial purposes (as opposed to engaging in commercial whale hunting), people were, generally speaking, not happy about a whale being hunted. Thus, protestors flocked to the reservation from all over the Pacific Northwest and the West and even from the East—people who were passionate about saving whales.

It's a long story (I wrote a book about it), but suffice it to say that I got to wondering why people would go out of their way to save a whale—especially in light of the fact that there was a lot that a little activism could have done on that particular reservation, a casino-free place with 75 percent unemployment at the time, with drug problems, with alcohol problems, with children sometimes unable to get medicine.

---

recreational boaters. Moreover, ecotours are offered of the Meadowlands area. They tend to stress the classical nature that is abundant in the area (and threatened by development). But tours of garbage hills are a more difficult sell than tours of beautiful, pristine-seeming green estuaries, just off the myriad of interstate highways, in the midst of the busiest travel corridor in the world, on the edge of the most densely populated city in the nation.

Not that it wasn't a beautiful place, with plenty of self-sufficient families carving out a rural life on the edge of the world, or at least on the gloriously beautiful basalt-cliffed edge of the continent that looked like the edge of the world, but it seemed as if people were quick to ignore human questions in an effort to consider the whale.

I began to study the history of whale-human relations in the United States; I read about the transformation of the United States from a whaling nation (whaling was our first global business) to a nation that hired artists to paint giant murals of whales on the sides of buildings in our cities, even in our land-locked cities, places where no whales had really been before. Whales went from being the thing that supplied the oil for your lamp, to being a kind of Platonic representative of watery nature. This led me in turn to ponder the following question: Whereas people would go out of their way to save a whale, what creature would they not care to save? What mammal would garner no protestors driving days from California? The answer, not surprisingly, was the rat. This led me to spend a year in an alley in a big city looking at lots and lots of rats.

I went to the city to live with men, a lot of men, as well as women, in particular my wife, and, as I said, with rats, which brings me to Henry David Thoreau, no offense to him. Thoreau, as naturalists everywhere know, built a little cabin on the outskirts of Concord, Massachusetts, and lived, in his words, "[a]lone and in the woods, a mile from any neighbor"—and ended up with a book called *Walden*. It is a book that is often thought of when people think of nature, especially (and almost exclusively) the non-dark side of nature. It is also a book that is, it seems pretty clear to me, often misread entirely. In the popular conception of Thoreau's "experiment," the bookish recluse moves out of the city, in this case the town of Concord. Thus ensconced, he meditates on nature, on pure wilderness, which, as the misreading goes, he believes he is in, which, in the end, is a good thing, a life uncontaminated and refined.

On the most mundane level, this misreading results in complaints about the legitimacy of Thoreau's experiment, in "gotchas." He often walked into town, people say. His sisters brought dinners to the cabin! He said nice things about the telegraph wire ("It was the sound of a far-off glorious life, a supernatural life, which came down to us, and vibrated the lattice-work of this life of ours"), and he took the train—he didn't *really* live in the wilderness! The gotchas are all the more upsetting in that it doesn't take much of an inclination toward metaphor to see that Thoreau wasn't trying to be so literal. He was, first, responding in part to the experimental communities that were in vogue at the time—

Fruitlands being the example established by Thoreau's friend Bronson Alcott. Alcott wanted Thoreau for his utopia. (If I had a utopia I would want Thoreau too, for, if he could be a little ornery, an enthusiastic arguer for argument's sake, he could also be very handy, as a carpenter, as a surveyor, as a lot of different things; he made his living as a kind of esoteric handyman, a philosopher for hire.) In the year before he built his cabin, Thoreau wrote in his journal, "We must first succeed alone, that we may enjoy our success together."

Walden can't be summed up in a sentence, but it can't be summed up as a nature book, either. It is a book that suggests we think about *how* we live, as individuals in society, in civilization, that we consider our nature and the nature of life among people. It is a myth, of sorts, that ends at dawn, at the beginning of a new life, with a picture of metamorphoses, the egg of a moth, "buried for ages . . . in the dead dry life of society," that might alight, that may be born, that, in Thoreau's words, "may unexpectedly come forth from amidst society's most trivial and hand-selled furniture, to enjoy its perfect summer state at last!" It's about re-examination and renewal, a myth designed to transform the hearer as he returns to live back in town.

Nonetheless, you find *Walden* in the nature section of your local bookstore. You see quotes from therein sprinkled like apples seeds around the world; you'd think Thoreau went to his desk each day thinking up pithy tree-loving aphorisms for the sides of coffee cups and driftwood etchings. Forget that he published a treatise on forest management; he is perceived as the poet of nature as a private place, as the mystical away. His most famous phrase—"In Wilderness is the preservation of the World"—is something that was published posthumously in his essay "Walking." He meant it, but it has taken on a life of its own, such that in wilderness is the misunderstanding of Thoreau, and, thus, wilderness.

Walden was accidentally lumped into the anti-urban strain of American thought—a strain championed very early on by Thomas Jefferson, the first great American anti-urbanist.[4] Jefferson could not bear the thought of the factory-driven masses living crammed together in the

[4] Anti-urbanism even continues today in a period of pro-urbanism, when, as a recent business news story noted, malls are decorated as if they were nineteenth-century city streets, with faux brick faces, etc. In the cities, meanwhile, there is a revival of downtown living. Ironically, though, many American cities are being rebuilt with urban-style housing that relies heavily on the accoutrements of car-oriented suburbia. Huge industrial lofts are retrofitted into vast flats of luxury housing, denying the city what has over the course of its history led to squalor but at other times to an efficiency of density.

cities; he thought of the cities as "sores." "The country produces more virtuous citizens," he wrote, though, as the writer J. B. Jackson has pointed out, when Jefferson referred to the *country* he was not arguing for rural solitude so much as for an agrarian society. Jefferson considered rural society to be more social, or, to use the loaded word, more *natural*. The misread Thoreau seems to fit right in here; he has often been quoted so as to seem to dislike the farmer. "[T]he farmer leads the meanest of lives," he wrote. "He knows Nature but as a robber." In fact, Thoreau's journals show him as a pupil of the farmer, and vice versa. For Thoreau, the farmer is a great repository of local history, of land knowledge, a knowledge surpassed only by that of the Native American—whom Thoreau, to the derision of his contemporaries, continually sought out—or that of the occasional ne'er-do-well, of whom Thoreau was likewise a great admirer, also to his neighbors disdain.

The idea Thoreau is exhorting here is that modern society tends to get in the way of man's primal relationship with the land, with the sacred everything in the soil, air, and water around him. He makes the "robber" remark in *Walden*, in his satirical farming chapter, to make the point that we are led purely by our desire for possession; in contrast, he himself cherishes enjoyment, which requires little in the way of materials or even beans. He saw as his great achievement that he needed little to enjoy life, despite people's grumblings about his not applying himself to anything worthwhile. (Even his neighbors did not seem to notice that he worked hard in the family pencil-making business.) Again, Thoreau is stepping back from society so as to consider how society operates and how a thinking person ought to operate in what might be a thinking person's society. *Walden* is a utopia of the "I," rather than a communal utopia. At the same time, it should be emphasized that the "I" of *Walden* is never trivial, never solely selfish. Bad nature writing suffers from a misreading of Thoreau's "I," confusing the Transcendental "I" with the personal self, the result being nature trivialized, merely noted. To build a cabin was the idea of Thoreau's friend, Ellery Channing, who suggested he set up a cabin on Ralph Waldo Emerson's wood lot, near Walden Pond. It was a place that was apart and still part of Concord, which is how he liked to be. (When Thoreau lived in the city, for a short time in 1843, he was unhappy, but mostly, it seems, because he missed home; it was a case of a fish out of the water—like James Joyce writing about Brooklyn, or Jimmy Breslin attempting to set up shop in Geneva or Lausanne.) I've always felt that Thoreau's romanticism—or the misreading of it—leads to the nature of the camping-gear catalog, of the Outdoor Channel, of wilderness writers opining about the wilderness, which is like shooting fish in a barrel. The misguided Thoreauvian keeps nature separate from everything else, even though everything else more and more threatens the very existence of nature. He or she sets up a divide between humans and nature that is unnecessary and mostly serves to hurt both sides.

Painting by Charlotte Hildebrand

To be sure, Thoreau was not in love with the city ("mere herds of men" is how he describes New York City in a letter to Emerson), but whereas Thoreau almost accidentally set up the dark side of nature, Jefferson intentionally engineered it, laying his conception of agricultural utopia down on the country, defining the national organization. More specifically, Jefferson engineered the National Survey of 1785, which in turn produced the grid, the criss-crossed dissection of the United States into uniform lots that still exists today, that makes us one of the only checkerboard nations in the world, an Enlightenment-era dream that, as you can tell from maps and cross-continental airline trips, actually came to be. The grid is, J. B. Jackson says, "the symbol of an agrarian utopia composed of a democratic society of small landowners." It's an apportionment of the land, with no regard, in fact, for the land.

Thoreau's romanticism led to the opposite of a square in the Jeffersonian grid, or to a pause in its relentless advance—that is, the wilderness area, the place that is *not* the rest of America, that is not man-infested: those John Muir–established sanctuaries where we can hope to restore ourselves, where we are, in a sense, in nature. In this way, wilderness areas are like parks and lawns and even the first suburbs, which (before their convenience to highways and discount big-box stores was advertised) were planned and green-manicured attempts to return city-focused men and women to a life communing with nature.

There is no place for the city in the grid, nor in the wilderness. In both cases, nature is *not* the city. It is its opposite. The city and nature, by our national geographic accounting, cannot co-exist.

But rats, when examined closely (though not that closely, of course), prove that nature and the city do co-exist, and that, really, the dark side isn't so dark. To my mind—and granted, it is a mind that has perhaps spent too much time pondering rats—rats prove that nature is not out there but right here, exactly where men and women live, even if a lot of naturalists don't want to think that. Once, I perused a giant and beautiful book of lush and exotic wildlife photos—a kind of pornographic nature book. Its subject was mammals, and it included the following line: "There comes a time when even the most energetic of animal lovers must part ways with the animal kingdom." This was proof to my mind that rats are considered to be apart. Thus, I went to the city to observe them, *á la* the naturalist, the patient observer of a nesting bird, only in this case the bird was a rat—again a reverse commute, and I, the commuter, hopeful for what I might notice passing me, going the other way. Each night in my rat year, I said good evening to my family, set out to a little alley, and sat on a camping stool and watched (through binoculars, through night-vision gear) as the tide of garbage came from the restaurants, the apartments, the fast-food stores, and as, in turn, the rats arose, ventured forth from their rat holes to eat, to fight, to flee, to gorge. I have too many observations to share (I wrote a book about them), but suffice it to say that the longer you watch rats, the more you learn about (1) rats and (2) the creatures that live with rats and that allow rats to live—that is, humans.

Rats, as rat experts point out, are commensal with humans—literally, they eat at the same table as people, the table being any place where there are a bunch of men and women eating food and leaving scraps, which is just about everywhere. Rats eat the same foods as humans (and even, according to some research, seem to especially *enjoy* the same foods—fatty and fried foods—as humans). The Norway or Brown rat, which, along with the Black rat, is the most prevalent in cities in America and in a good part of the world, exists only where there are humans present. Brown rats have many different common names, such as alley rat, city rat, farm rat, river rat. The most appropriate common name would be *human rat*, because if there are no humans, no human history in an area, then there are no rats. (I once read of a scientist who studied rats finding a colony of Norway rats and no humans on an island in the South Pacific, but then upon further investigation he determined that they lived in the area of what had been a garbage dump for a Japanese prisoner-of-war camp.) Rats live on farms and rats live in the suburbs and rats are easy to find out behind cute little country stores, but generally speaking, rats are an indicator of the presence of human settlement. If you put aside for a moment all the grizzly bears eating the trash at Yellowstone or Glacier National Park, then the presence of grizzly bears means an absence of man, a not-so-settled rural area. The presence of rats—the presence, in particular, of hundreds and hundreds of thousand of them—means *city*. The reason that rats are not invited into the pantheon of sexy mammals is that they live in our rejectamenta, in the foul stream of refuse that we tend to deemphasize when we think about nature and ourselves, even though we make it. That's our dark side, the part we prefer to ignore. We think that rats are the dark side, but garbage and waste is the dark side, or dark to us. It is also the natural habitat of rats.

Rats in their natural habitat: the idea itself is the bond between humans and *Rattus norvegicus*, and through this idea, through nature's dark side, I like to imagine humans marching triumphantly back into a new conception of nature, one that has no divide. Rats are the creatures that say that man and nature are not separate, that on the so-called dark side, the first thing you run into is the human. Of course, there are not a lot of people marching triumphantly. That's because to walk through the dark side you have to take the basement door, the hole in the cellar, the drainpipe under the world. You have to go into the tenements that still exist. You have to spend time in uncared-for parts of cities, in underserved neighborhoods and housing projects. You have to see where the rendered fat exits the Park Avenue restaurant, the way in which the SUV that you use to drive to the National Park trailhead is spilling nonpoint source pollution, its brakes excreting chemicals and metals into streams that are highway-side and thus on the dark side.

There's not a lot of interest in the so-called dark side, obviously. It's downplayed, naturally. We don't celebrate it. The result? We keep things separate. We don't see our actions, the *how* we live, as relating to Thoreau's nature; we see ourselves as minding our own business, our special spot in the grid. We see our lives as separate, as apart from nature, like rats.

There is no special place, of course; the tendency of nature writing to celebrate the extraordinary in the natural environment tends to make other places seems less than extraordinary, or bad—to rat-ify them. And that's not fair, and is even unjust, especially as applied to other species, like man, or rats. Exceptionalism is a form of blind boosterism, and it can lead most obviously to trash incinerators in low-income neighborhoods, but it can also lead to missed opportunities: the opportunity that windmills in New York harbor might afford, for instance, in the view of luxury condos. There are lots of good things about living in the country, lots of chances to observe easy-to-savor romantic nature moments. If there is a good thing about living in a city, from the point of view of the conception of nature in our day and age, it may have to do with the reverse-commute theory, for we are placed well to notice the things that others might take for granted. A hawk that lives on an apartment building is no more unusual than a hawk that lives on a cliff; it's the same hawk. It's not an extraordinary event for the hawk at all; it's a pragmatic solution. It is, though, an extraordinary opportunity for humans to see the human-built environment's effect on the non–human-built environment—and maybe, even more importantly, an opportunity to see the human-built environment as part of the environment. It is in our nature to build things, to discard them; we also have it in our nature to see ourselves building things and discarding them. We also have it in our nature to see ourselves. *Nature,* after all, comes from the Latin *natura,* meaning birth, and *nasci,* to be born. We are born into nature, into all, into the universe. All of us, rats included. The difference is in what we see, or in our *environment,* a word that comes from the Old French and means *all around,* as in everything, everywhere, as in seemingly high and seemingly low.

Seeing is what Thoreau cherished above all else. His last entry in his journal—a vast, two-million-word, twenty-five-year-long project—describes the effect of a violent rainstorm on some railroad-related gravel, and ends, "All this is perfectly distinct to an observant eye, and yet could easily pass unnoticed by most." Of all the disciples he has inspired, one of my favorites is Benton MacKaye, the cofounder, with Aldo Leopold, of the Wilderness Society. Leopold is well known for his concept of the land ethic. MacKaye is best known as the inventor of the Appalachian Trail, though the intention of the trail is also bastardized in most references. Yes, it is a hiking trail from Georgia to Maine, along the Appalachian Mountains. Yes, it is an outdoor recreation device, built, excitingly, by local groups working for a regional purpose. But MacKaye did not intend it merely as a hiking trail; he designed it after looking at the flow of human traffic, after observing the movement of natural resources in the region, after drawing the patterns of human settlement. If Leopold was a land philosopher of sorts (albeit a forester philosopher), MacKaye was the land engineer, and the Appalachian Trail was a piece of a planned redesign of how we live in the cities and even in the country. It was an answer, in part, to the question he posed in 1928, in his book *The New Exploration*, a question that still ought to be posed today: "Can we make of this time and century something better than a chaos of industrial cross-purposes?"

"We find ourselves in the shoes of our forefathers: their job was to unravel the wilderness of nature; ours is to unfold the wilderness of civilization," MacKaye went on to say. "Or are we to be lost in the jungle of industrialism? Are the elements of water and steam and fire to remain our masters, or will they become our 'servants for noble ends'? Are we going to ride on the railroad or let it ride on us?" As with Thoreau, the answer for MacKaye was in seeing and understanding the forces of society, the actions of the herds, and MacKaye saw the cultivation of observation as a noble and indispensable task. Here, in the closing of the same book, he describes in his somewhat over-noble language, the task, calling for "a new explorer":

> The new explorer, of this 'volcanic' country of America, must first of all be fit for all-round action: he must combine the engineer, the artist, and the military general. It is not for him to 'make the country,' but it is for him to know the country and the trenchant flows that are taking place upon it. He must not scheme; he must reveal: he must reveal so well the possibilities of A, B, C, and D that when E happens he can handle it. His job is not to wage war—nor stress an argument: it is to 'wage' a determined *visualization*. His attitude in this must be one not of frozen dogma or irritated tension, but of gentle and reposeful power: he must speak softly but carry a big map. He need not be a crank, he may not be a hero, but he must be a scout. His place is in the frontier—within life's 'cambium

> layer'—the fluid twilight zone of all creative action in which the flickering thoughts of future are woven in the structure of the past. . . . And our Last instruction to our new explorer and frontiersman is to hold ever in sight of his final goal—to reveal within our innate country, despite fogs and chaos of cacophonous mechanization *a land in which to live*—a symphonious environment of melody and mystery in which, throughout all ages, we shall 'learn to reawaken and keep ourselves awake, not by mechanical aids,' but by that 'infinite expectation of the dawn' which faces the horizon of an ever-widening vision.

When I was watching for rats, I had an infinite expectation of the dawn in the literal sense too. When you are watching for rats, diurnal creatures that they are, you are praying for the dawn, which is around the time that rats, which seem to be on actual time clocks, come up out of their holes for their last feedings, for their last shot at the dregs of the night, what the trash collectors missed. Of course, you also see humans. Coming into the alley at dusk on a fall evening, or on the subway back and forth, or riding my bike in the empty city streets on my way home around midnight, I saw all kinds of people, observing. I watched the people as well as the rats—noticed, for example, the people running in huge packs down into holes at night, watched them stumbling around in smaller groups in the streets late at night. I saw the herds walking the same paths in the mornings, saw them scattering bits and pieces of breakfast food, saw their bars and restaurants hauling out loads of discarded scraps. The garbage feeds the rats, of course, cultivates them—after a while, you begin to realize that cities seem to be raising them, like chickens or cows. The scene of the rats, after appalling me, led me to a sense of the urge to herd, to be, in Whitman's phrase, "one of the crowd." It led to a sadness, for various reasons that are probably obvious in part to anyone who has walked around alone in a big city at night, a sadness that, however, ultimately revives. In her poem "Wild Geese," Mary Oliver writes:

> Whoever you are, no matter how lonely,
> the world offers itself to your imagination,
> calls to you like wild geese, harsh and exciting—
> over and over announcing your place
> in the family of things.

In the course of my rat year, I also found out why people think there is one rat for every person. That idea is the misapplication of a nineteenth-century British estimate that put the rat population in England at one rat per acre—a statistic that transmogrified into the one-rat-per-person statistic that has been used variously by the U.S. federal government, by the United Nations, and by the city of New York (which once hosted a scientist who disproved it). The statistic survives because people want to believe there is a rat out there for every one of them. It takes some pressure off. You're not so bad, whoever you are, if there's a rat out there to compare yourself to. People know that there is a dark side. They just don't know that it's them, and that it's really not that dark. It's just human, or human nature, depending on how you look at it.

The Futures of New York

Anne Matthews

Taking New Jersey Transit from Princeton Junction to New York's Penn Station means a fifty-eight-minute trek (off-peak, one way) through a profoundly disturbed landscape of chemical mudflats and industrial slurb. Yet crossing the Meadowlands one bleak February morning, I saw from my commuter-train window a dozen egrets, flying fast and low, an arrow of white headed straight for midtown. What are they *doing* here? I wondered, horrified, amazed. How do they *live?*

Nature knew what I did not: that over thirty years of environmental cleanup have brought egret and bittern, glossy ibis and yellow-crowned night heron back to city waters. In the shallows of Jamaica Bay, on uninhabited islands in the waters off the Bronx and Queens, hundreds of wading birds now breed. To find these other New Yorkers, you may need to crawl ashore through great tangles of poison ivy, past rusted Chevys, then hold up a truck mirror to observe the secret rookeries—but they're there, and flourishing. I had no idea.

It was a figure-ground problem. For years, I had looked at Greater New York and seen only what I expected to see: a profoundly unnatural landscape; a competitive maze; a wonder of money and art that seemed a thrilling human triumph some days, and on others a declensionist disaster. New York City attracts jeremiads. Emerson called it a sucked orange. Fitzgerald pronounced its grimy suburban sprawl "the ugliest country in the world." Vonnegut saw Manhattan as our skyscraper national park. Yet above, around, behind, below, I began to discover another city, suppressed and segregated during daylight, exceedingly lively from twilight to dawn. And I began to wonder what the future might hold, for both New Yorks, since so many environments collide in the five boroughs—northern and southern climate zones overlapping, salt water mixing with fresh, land melting into ocean—and yet, of all U.S. cities, nature and culture here seem most spectacularly, insistently estranged.

The key word here is *seem*. The city is strikingly wilder today than in 1900, or 1950, or even 1980. The Bronx has a healthy coyote population now, urban pioneers from Westchester. Porpoises play again in the Hudson. Wild turkeys are colonizing not only Riverside Park but Central Park; apparently they fly down Broadway late at night, then take a left at Lincoln Center. The blue crab and fiddler crab populations are up, way up, scuttling in their millions along the silty floor of a re-oyxgenating harbor and estuary. Deer have come back to upper Manhattan, making late-night forays along the Amtrak trestle. And black bear have been exploring the Palisades Parkway, *and* Chappaqua, *and* the dumpster behind the White Plains Bloomingdale's. To the familiar calendar of New York events, from Opening Day to Marathon to Tree Lighting, we should, perhaps, add other mileposts. Like mid-April, when peregrine chicks hatch atop the Throgs Neck and Brooklyn bridges, joining the world's largest urban falcon population. Or the full-moon nights of June, when thousands of horseshoe crabs come back from the Atlantic deeps to mate along the Brooklyn shore, as they have done since dinosaurs roamed New Jersey. August is, reliably, monarch time in the city, when clouds of butterflies commute to Mexico by way of Fifth Avenue; September, in this other New York, is moving season. Fifty-pound striped bass return to the Hudson from a summer in the Hamptons; urban rat packs migrate from Central Park to the Upper East Side. And even in December, you can sometimes stand on Wall Street in the small hours and hear birdsong, faint and high; migrating birds pass over the city nearly every night of the year.

Wildlife biologists make it clear that the resurgence of nature in and around New York City is no anomaly. Clean air and clean water legislation from the Nixon administration on, intensified wildlife-restoration programs, and assorted hunting and fishing bans have all helped U.S. animal populations soar just as a matching development boom, over the last ten years especially, has brought an extreme expansion of the built environment: 80 percent of all structures in the U.S. today were built since 1950. Forcing Colorado condos deep into elk country, planting Los Angeles neighborhoods in traditional mountain-lion terrain, dropping New Jersey exurbs into deer lands, shoving Florida golf courses into crocodile and alligator habitats. As the continent's dominant species, we tend to forget how closely other creatures observe our works and days: our high-speed roads, our mystifying preference for sunlight, our magnificent garbage stream.

So: two population jumps, one crowded nation; one extremely crowded tristate region. Many species flee us, or try to. But others, plant and animal, adapt to newly humanized landscapes with terrible patience—and discover, in the process, that city and suburban life can be far less stressful than a career in the wild. To a fox, or a falcon, our dense complex of urban greenways, corporate ponds, and delectable garbage can look remarkably like a giant animal sanctuary, where the takeout is superb, and the only predator is the SUV. When cultural carrying capacity is exceeded—when humans stop saying *awww* and start dialing 911—that is when our easy assumptions about what is urban become strained, and the real dilemmas surface. For human and nonhuman to covet the same real estate is no light matter. New Yorkers like to say they're all for diversity. But in a town that traditionally prefers its nature in a museum diorama, or else on a plate, *bio*diversity can be hard to sell.

My turf, *my* place, *my* habitat, *my* real estate, *my* home. Humans are often so self-involved, so anthropocentric, that the other, quieter urban immigrations of our time are easy to overlook. And never have our responsibilities as top predator, ethical and practical, been more unclear. Already, one American in twenty lives in New York City, or in the city's suburbs, as half of humanity now dwells in the world's metropolitan areas. From Toronto to Tokyo, confrontation and competition with the natural is becoming part of city, suburban, and periurban routine. Some encounters charm us, some we dread, others we badly misunderstand, which is unfortunate, since history and archaeology tell us that messing with the natural world generally hands an urban culture one of three outcomes: an altered life, a lesser life, or a long night.

To learn how and why nature returns to the city, I've talked with scientists, planners, officials, and historians, accompanying many such experts on their forays into the urban wild. When discussing New York's natural present, they were eloquent; when considering New York's environmental history, the city as eco-artifact, wonderfully informed. But on the subject of the city's future, such conversations often turned edgy, anxious, doubt-filled. What *will* the New York of our children and grandchildren be like, forty or fifty years hence? Fifty years is a hard span for the urban imagination: near enough to be personal, distant enough to baffle, or seduce. The many bad guesses of the 1950s remain visible from almost any window in America. By 2050, the U.S. population may approach five hundred million. One-quarter of us will be Hispanic. Almost one-fourth of Americans will be over sixty-five. And in some ways, New York 2050 is already here. Its shade trees are our garden-center saplings, and almost a million 1950 babies will live to sit under them as centenarians.

Some imagined urban futures are reassuring, others dire, but nearly all evoke an ancient dread: nature coming back to a human-claimed place, the return, the retaking. If one surveys recent studies and models produced in the last ten years under the auspices of (among other institutions) Princeton, Columbia, NYU, CUNY, the United Nations, the U.S. Congress, and the New York Academy of Sciences, it may be possible to assemble at least a tentative view of four futures that may await New York, none definitive, all plausible.

Future #1 is a dystopian battleground of civic chaos and decay, a *Bladerunner* vision of New York in which the megalopolis continues to expand without constraint. The result: a debilitating entropy marked by worn-out infrastructure, vast squatter settlements, fraying public and social services, and battered ecosystems. (For a glimpse of such conditions in real time, try some of New York's sister megacities, like Mumbai, Dhaka, Lagos, Sao Paulo, or Jakarta.)

A second, milder edition of this chaotic future is New York as a deindustrializing, formless city, hit hard by corporate downsizing and job exports—a Greater New York that retains a vigorous tourist city at its core but is otherwise a borderless metropolis, fully absorbed into the Boston-to-Norfolk eastern megaplex, and driven by the anywhere/anytime interactions of infotech and e-commerce. Already, geographers like to call New York a galactic city, an ever-expanding tissue of urban development that fragments natural habitat and engulfs once-freestanding towns and regions, such as Princeton, New Jersey, where New York and Philadelphia have finally touched, or Lancaster County, Pennsylvania, 125 miles from Manhattan, now clearly the edge of New York commuter country. By 2050, for the first time in four centuries, Americans may be unable to walk away from a fouled urban nest, especially as our population tops four hundred million. Hope for ingenious retrofitting and liberation from fossil fuels; expect extraordinarily painful debate on what cities should do and be in a built-out postindustrial information nation.

Vision #3 for a future New York is the deindustrializing city morphed into a sustainable one, where a new spirit of urban ecology plus permanent energy woes persuade us not to merely use and abandon urban places, but to use and reuse them. Infill plus green initiatives are the common imaginative thread here, a wholesale embrace by New Yorkers of what is already fact elsewhere: urban farming in Philadephia and Trenton; Chicago's rooftop gardens; Paris's huge new investment in city trees; Salt Lake City's encouragement of reflective roofing to cool the urban heat island. More light rail, more public green space, London-style congestion fees to discourage driving in midtown look more alluring every year. Change *is* possible: in 2005, for example, after decades of self-interested foot-dragging, Manhattan building owners finally agreed to dim high-rise lights during peak bird-migration season, as Toronto and Chicago have long done.

If New York can revamp its skyline habits to let songbirds pass unharmed, why not restore other aspects of the city's original nature, the ultimate environmental reparation? New York, after all, specializes in creative destruction: building up, tearing down, rebuilding some more. Why *not* steam-clean the harbor, reseed the oysters, replace invasive phragmites with native saltgrass? Environmental historians, wildlife biologists, politicians and urban experts all hyperventilate at the prospect, for different reasons, as arguments for ecorecovery pile up; interpretations turn ever more fluid and contrary, though always with a subtext of uncertainty, of doubt.

Which New York do we restore? The pre-European-contact ecosystem of 1500—or the islands Peter Stuyvesant knew, steadily morphing into a pastoral landscape of village and farm? Do we make our restoration set-point 1900? Or maybe 1950? And all the while, various city, state, and nonprofit groups attempt a stubborn, visionary piecework, public and private, of native-species restoration—blight-resistant chestnuts given sanctuary in Central Park, pearly everlasting reintroduced on Staten Island, bobwhite quail released in the Bronx. What *is* wild, these days? Maybe we should look for new definitions of wildness between the cracks of a Brooklyn sidewalk, in a neighbor's affection for plastic lawn flamingos, in a North Jersey swamp filled with broken classical columns from the old Penn Station, or even in our own turbulent bloodstreams.

Defining natural is even harder, since even designated North American wildernesses are very large parks in disguise, with rangers, permits, and posted trails; only a handful of Alaska's mountains and Nevada's alkali plateaus remain officially unmapped. Ought we be calling these few untouched environments *first nature* and everything else *second nature*, as the British cultural critic Raymond Williams proposes? Or, instead of the long-running either/or standoff of front lawn versus wilderness, would it be more grown up to make the garden our model instead? Even a megacity, scholars now suggest, is part of a larger land-use story. Cities may be as vulnerable to nature and fortune as any other life form; some endure, some thrive, some shrink. Some vanish. But other critics are repelled by talk of the lifespan of cities, or the decay and regeneration of suburbs. Cities are artifacts, they insist, designed to abstract nature, at a profit, over a distance, turning cows into hamburger and redwoods into plywood.

New York's newly wild nights, I would argue, mark a genuine change in city-country relations, the first in centuries. For four hundred years, New York has resolutely kept moving, to find the best deal, to not waste time, to never show weakness. It has rarely looked around, rarely looked back. Maybe it should. Wild does not always mean natural; urban is not the same as tame. Even in Manhattan, you are never more than three feet from a spider.

And urban vision #4? That would be New York as a coastal megacity profoundly altered by global warming. In the wake of Hurricane Katrina, New York officials rightly worry about the effects of giant storms in particular, and global warming in general. In the new century, it seems, we may occupy a noticeably hotter, wetter New York, with increasingly dramatic weather—at least two big hurricanes a year, plus more blizzards, ice storms, hailstorms, and storm surges. "The Baked Apple: Metropolitan New York in the Greenhouse," a 1996 report by the New York Academy of Sciences, still makes persuasive reading, given its vision of a New York in 2050 with far more ozone, more smog, more power shortages, more asthma cases, and more heat deaths, as New York summers begin in April and last through October.

New York is, above all, an urban archipelago, a port city seamed with rivers, subways, and sewage-treatment systems. Even if global warming creates only moderate sea-level rises, it is entirely possible that by 2050, Battery Park City, the Rockaways, Coney Island, Alphabet City, Red Hook, Jersey City, and much of the financial district will all be drowned land. The F.D.R. Drive and the Gowanus Expressway may need to be shored up, as the ground beneath them takes on water and turns unstable. If the tide gates at city sewage-treatment plants are submerged (as already happens during especially hard rains), raw sewage will assuredly back up into New York streets. La Guardia Airport, only six feet above sea level, will eventually have to close, and Newark too; dikes may keep Kennedy open for a few decades before it too succumbs. In this mildewed metropolis, Lower Manhattan's 4, 5 and 6 trains will rarely be usable. City curfews may become common, as repair crews work by night to clean up storm damage. Parts of the West Side Highway will almost certainly be under water; much of Greenwich Village, distinctly soggy. The beachfront Hamptons, like Nantucket, Martha's Vineyard, and the Jersey Shore, will be only a memory. And expect Central Park and Prospect Park to be largely junk trees and weedy species; this overheated New York is no place for oaks, or wildflowers, or songbirds, though it delights coyotes, Formosan termites, mosquitos, and roaches. *Lots* of roaches.

Speculation? Yes. Science fiction? Not any more. It's unlikely that any of these urban futures will come precisely to pass. But we *can* know two things about Greater New York five decades hence. First, if the last fifty years are any guide, most of the stresses, dreams, disasters, and component parts of 2050 are already with us—unrecorded, overlooked, misinterpreted, buried on page D24. And for that reason, scientists, citizens, academics, and the media all need to make New York's culture/nature tensions a significant part of the public conversation to come, and need to accept, however ruefully, that nature, in the long run, considers us lunch.

Compromise has never been the New York way. Faced with forest, desert, prairie, river, humans tend to make a space in the midst of, force a path through, beat back. But the price of dominion still troubles us, if only in poetry, or dreams: the lone and level sands stretching away from the toppled statue; the great globe dissolving into illusion; the white whale, hunted, that turns to hunt us; just as the Biblical text least likely to be quoted by tristate developers is still Isaiah's unsettling vision of the heron in the living room. ("Woe unto them that join house to house, that lay field to field, till there be no place that they may be alone in the midst of the earth! . . . Yet soon they will call, and none shall be there; thorns will come up in the palaces, nettles and brambles in the fortresses; it shall be a habitation of dragons, a home for the bittern.")

Bottom line: anyone, anywhere, desiring a livable future for an urbanized planet must still look to New York, where the nature/culture contrasts are extreme but the ecological footprint remarkably small. This is a megacity that stacks and packs its residents, then makes them use mass transit; a megacity where nature's advocates are learning, more and more, to insist that environmental issues are not an aesthetic luxury but a core health/family/economy voter issue. New York's trademark knowingness and tenacity, its sheer pleasure in argument, may be the saving of us all.

But only maybe. What you see, as ever, depends on where you stand. And after my sojourn on New York's wild side, here is what *I* see, each time I leave the city on the commuters' local.

The tracks south and west out of the city follow the old Pennsy route, dipping under Hell's Kitchen and the Hudson, then running in darkness until we burst into a Meadowlands doubly gilded. A crimson sun is sinking toward New Jersey's Kittatinny range, first cousin to Central Park's lost Alps, even as a full silver moon—the hunter's moon, New York colonists would have called it—soars above the asbestos outcrops of Staten Island. The Hudson rail tunnel is almost a century old now, one track in, one track out, and to let an express go by my commuter train stops not ten yards from the sleeping rubble of another New York. Under the eighteen-wheelers in Secaucus trucking lots, earth and water and toxins and time are persuading Bronx tenement bricks into mud and dust once more, grinding glass from a prewar Queens apartment again into sand, blurring chiseled Manhattan stone to elemental granite. You cannot step in the same city twice. Beside the train tracks, a stand of native saltgrass bends above water dark as peat and bright as pewter. I watch the last light touch and leave and touch again the city skyline and I know (although I cannot see) the wild night beginning there: the commuter rats trotting blithely to work, the Wall Street peregrines gliding toward the Hudson, the midtown skyscrapers swaying in the May wind like giant trees, the horseshoe crabs of the continental shelf turning once more toward the Brooklyn shore, the Bronx coyotes watching the great bridges and waiting for dark. The train begins to move again, across the Secaucus waterlands, and as we turn west toward Newark I cling to the swaying leather seat and look toward Manhattan once more, but the twilight is tricky today. I see only a tangled bank; a green cliff above the estuary; a glitter of minerals on a rising sea.

Imagination, Beauty, and the Urban Land Ethic

Teaching Environmental Literature in New York City

Devin Zuber

The problem with nature in New York is that there isn't any.

A student said this to me in a class I was teaching, Environmental Literature, in response to a question I had asked about how one can maintain contact with nature and the wild in New York. Her response was typical of an attitude I encounter at the start of the semester in this course: a prevailing assumption that nature is separate from our densely populated streets, that it is in upstate New York's Adirondacks, in New Jersey's Pine Barrens, perhaps, maybe, in the north of Central Park, but certainly not in the urban spaces we inhabit as New Yorkers.

Teaching literature with an environmental edge, engaging urban students with the growing field of what English professors are calling *ecocriticism*, always brings me back to some core questions about the discipline when I face the misperception that nature is "out there" and beyond, but never in our own city backyards: What role is literature to play in fostering a broader environmental awareness, especially in a multicultural city of millions? How does reading poetry and the process of writing relate to the beautiful, complex worlds outside of the book and away from the classroom?

I believe environmental literature has a special function in nurturing what Aldo Leopold famously termed the *land ethic*: a deepening connection to the areas we inhabit, an awareness of the systems that shape our habitats into something unique. "This," in Leopold's words from *A Sand County Almanac*, "reflects the existence of an ecological conscience, and this in turn reflects a conviction of individual responsibility for the health of the land." The land ethic is thus a moral imperative that is contingent on our perception; it transforms the abstractness of living in space into a concrete and localized place. In William Kornblum's essay that appears in this volume, he describes how simply walking around the local streets with an attentive eye is the first step in fomenting a land ethic in his students. Reading natural histories of local places also certainly helps, and Kornblum and many others in this book have made important contributions to broadening New Yorkers' collective sense of place. (After finishing Sullivan's recent *Rats*, I must even admit a new regard for the millions of hardy vermin that crawl under our streets and through the walls; see his essay here for an account of that book's process.) As an English professor and not a natural historian, I would further argue that the land ethic is fundamentally enriched by literature of the imagination, by so-called fiction, and that such work should be a core part of any environmental literary curriculum. If a land ethic is contingent on the senses that observe the areas we inhabit, and these observations are enriched by knowledge of ecosystems and circulations often invisible to the naked eye, it is equally dependent on the human imagination: the universal and innate capacity to image ourselves and others beyond the confines of what is seen, heard, and felt. Poetry feeds the imagination, which in turn energizes our perception and broadens the space within us for a potential urban land ethic.

Ever since the Enlightenment of the eighteenth century, with its cult of reason and embrace of empirical science, imagination has often received short shrift, paling alongside the virtues of logic and universal objectivity. The Romantics rebelled against this paradigm and struggled to bring back an appreciation of imagination as an intrinsic part of human perception. Not coincidentally, many have argued that in poetry by William Wordsworth, Samuel Taylor Coleridge, and William Blake (among other Romantics) we can locate the first glimmerings of an ecological consciousness, a green thinking that anticipates our modern conservation concerns. Still, our dominant contemporary world view remains close to the secular humanism that emerged in post-Renaissance Europe, with the imagination viewed, at best, as a sort of entertainment luxury, and at worst as a dangerous source of delusional fantasy. I often point out to my students a relationship between the histories of poetry and environmental crisis in the twentieth century: that precisely as poetry declines as a popular mode of cultural expression, environmental problems become more endemic, and are related to a growing perceptual disconnect from nature. Both facts are obviously part of a larger story of industrialization and growing technologized artifice; nor do I mean to imply that a culture with great poetry necessarily has a strong land ethic (although one could certainly point out the many oracular, indigenous peoples who are keenly attuned to the nature they live in—certain Aboriginal tribes, for example, literally sing their landscape to life with songs that teach important values about the ecosystems on which they depend). I do think, however, that the disconnect New Yorkers can have from the spectacular nature right under their noses is very much a failure of the imagination, and poetry can work to redeem this capacity.

Christopher Meier and Tony Hiss's work on the geohistorical layers of New York illuminates the role of the imagination here. Meier and Hiss peel back the topographies of the city, showing some of the earliest European cartographies that shaped Manhattan streets, and then go back further, to the glaciers and land formations that caused the confluence of the Hudson and the Atlantic. They write of this geologic knowledge as our "second address," which we all have a right to know, and point out that awareness of our second address explicitly requires the imagination. To truly grasp the levels beneath our feet, we must step beyond the visible, sometimes going back in time millions of years to events that are still determining how we live in this vast watershed today. The power of the imagination in grasping the scope of these time frames, what scientists often call *deep time*, has long been realized by innovative pioneers usually associated with the most rational developments of scientific thought. As Charles Darwin sailed on the *Beagle* around South America in the 1830s, trying to piece together the gleams and insights he had gained from the forms of animals on the Galapagos and elsewhere, he struggled to find the language for a theory of evolution that had never been articulated before. While it is widely acknowledged that Darwin was reading much of the work of geologist Charles Lyell during this formative period, it is less well known that Darwin also repeatedly returned to the difficult language of John Milton's *Paradise Lost* (1667), finding in that imaginative poem about Adam and Eve something that helped shape out his ideas—ideas that are so intrinsic to our thinking today that we hardly register their strange, radical nature; ideas that overthrew nineteenth-century paradigms about the origins of life.

The need for the scientist to drink from poetry, finding nourishment for his or her thought in words that stretch the imagination, continued into the twentieth century. "Imagination is more important than knowledge," Einstein once said; "Knowledge is limited. Imagination encircles the world." When nuclear physicists after Einstein began naming the subatomic particles that new technology had made apparent, they turned to James Joyce's *Finnegans Wake* (1939), finding their "quarks" in Joyce's whimsical prose. More recently, proponents of chaos theory have found the early Romantic work of William Blake to anticipate some of their dynamic flow models. Blake's attack on Newtonian science, which saw nature as a wind-up clockwork machine, has caused him to be the poet perhaps most quoted by contemporary scientists working on unified field theory.

It is useful to draw such connections in an environmental literature course that tends to attract a fair share of students not majoring in English, most often from the earth sciences, students who are better acquainted with Albert Einstein than with William Blake or Wordsworth. In an era when environmental issues are increasingly transnational and require initiatives that cross borders (global warming being the most obvious of these), a course on environmental literature can productively operate on a microlevel to build interdisciplinarity and create affiliations between fields of science and literature that are usually seen as mutually exclusive, even antagonistic to each other. This pragmatically begins not with lecturing on Darwin's reading of Milton, or what Einstein has to say about the imagination, but with the basic activity of connecting the words we read on the page to the world we observe outside, the world that we collectively inhabit as New Yorkers.

The student writings this simple process has inspired continue to impress me, and they have broadened my own land ethic as a professor working at Queens College. One student's reflections on a poem by Wordsworth that dealt with birds led to an analysis of the avian wildlife on the campus, and the delightful revelation (for me) that a pair of red-tailed hawks roost atop the tallest building on campus—their presence explaining, suddenly, the relatively low population of squirrels on the campus grounds, which I had often wondered at. Many students have written eloquently of what it was like during the infamous blackout of 2003: the stars that came out, reminding them of the deep sky that was always there but rarely seen, the leisurely walks on an evening normally spent in the air-conditioned indoors, bringing renewed awareness of the summer songbirds who call the night in.

As the students work to uncover Queens College's "second address," using their writing to perform a kind of archeology that is both ecological and cultural, someone will inevitably unearth the fact that Walt Whitman once taught in a tiny schoolhouse on the land that subsequently became the college (though only after other layered histories of the space; it was an orchard farm, then an all-boys reform school for the "truant and delinquent lads" of Queens). That Whitman was once literally present on the ground on which an environmental class now takes place couldn't be a more felicitous accident of space and time.

Unlike many other Romantic poets (both European and American) who were entrenched in the city-as-bad, country-as-good dialectic, Whitman fully embraced the contradictions and whirling chaos of New York City. The free verse of his *Leaves of Grass* (1855) forever changed how modern poetry could be written, and some ecocritics (literary theorists who explore the ecological and environmental implications of texts) have found Whitman's long-flowing language to operate as a biological system unto itself. Whitman's words, they argue, incorporate dynamic flows of vernacular language, creating organic and uneven structures of thought that shrink and expand in asymmetric patterns just like the pulses and flows of an ecosystem that holistically work to unify the disparate parts of a whole. Thinking about Whitman in this way can be a little abstruse for the undergraduate classroom, but students certainly find Whitman's work on New York City readily speaking to them some one hundred and fifty years after it was written. Mass transit in the city still shows the dazzling variety of the human species, as it did for Whitman's anthropological eye in his "Crossing Brooklyn Ferry," which remains one of the greatest odes to public transportation: "Crowds of men and women attired in the usual costumes! how curious you are to me! . . . the hundreds and hundreds that cross, returning home, are more curious to me than you suppose."

Whitman's free-form style and careful observations of New York topography can teach students how to keenly observe their own fellow creatures interacting with local environments, and how to creatively render their thoughts without the limitations of form or rhyme. Often the experiences that students write about as they link poetry to place return—just as they did for Whitman, again and again—to an intensely personal encounter with beauty. This contact with the beautiful, it seems to me, is a critical resting point for the imagination, and is perhaps the vital juncture between an experience in a poem—of reading words that make one feel something—and a feeling in the sensuous world of perception outside the body, in the world for which the land ethic tries to cultivate a moral responsibility. The idea of beauty has been vigorously interrogated, critiqued, and problematized for the last one hundred years (certainly as much as the imagination has come under fire), and I do not wish to stray here into the cross-fire of competing scholarly debates. By an encounter with *beauty* I simply mean an affective experience where one is emotionally moved (even to tears) by something larger and other than the self. In *On Beauty and Being Just* the philosopher Elaine Scarry has written of how beauty radically de-centers our sense of who we are, making us humbly open to things not normally in our purview; beautiful things tear open the world "and pull us through to some vaster space."

A challenge for the teacher of an environmental literature course is thus getting students to be open to beauty in their own neighborhoods, and not to think, again, that beauty is only available in vast, rural spaces such as the Adirondacks. If anything, the dense urbanity of New York City offers an exemplary model for undoing the human-versus-nonhuman dichotomy that underwrote the creation of the National Park system and that informs our more dangerous technological arrogance of the twenty-first century. New York lets us see instead the human as intrinsically part of vast, single interdependent web. In her essay in this collection, Anne Matthews cites several well-known assessments that predict that the next century will be increasingly, if not relentlessly, urban. This need not be a dystopic vision, as Matthews makes clear—one of the four futures she postulates for New York sees it as a city healthily deindustrialized, with accessible area waterfronts, increased greenery, and sustainable local economies. As part of the process of getting there, it seems we must embrace the contradictions of an industrialized city-in-nature, and nature-in-city, as Robert Sullivan and also Hiss and Meier emphasize in their essays when they rhapsodize about New Jersey's Meadowlands; and moreover, I think, we must seek to find beauty in such unlikely conjunctions.

One project I sometimes assign reflects on the aesthetics of paintings by Caspar David Friedrich (1774–1840), and then asks the students to locate a place in the city that refracts, somehow, a similar ethos about nature. This seems at first wholly contradictory; we study Friedrich's work, perhaps the most quintessential landscape paintings of the Romantic period, when we cover the visual culture of the early nineteenth century. His work is iconic—isolated figures on mountaintops and deserted beaches that typify a solitary communion with nature as spirit—and his paintings are undeniably, exquisitely beautiful, even as they also present certain transcendental dangers of Romanticism (going beyond and above nature). The students' challenge is to capture, with photography and words, a place in New York City that offers a similar contact with the beautiful—not in spite of the industrialization and flow of multitudes of people that are so absent in Friedrich, but precisely because of these modern factors. In a superb example, a student wrote how the public promenade of the Brooklyn waterfront, captured in the accompanying photograph, works to divide the space into horizons, much as in Friedrich's *Evening (Sunset)*; and it was important for her to include figures that were absorbed in the beauty of the moment, just as are the figures in the nineteenth-century painting. In a wonderful modification of Friedrich's anthropocentrism, her photograph replaces his dominating figure of the human, silhouetted by radiant light, with a tree, so that the tree, too, appears to participate in the moment's beauty. The Brooklyn locale had a further, personal poignance for this student, embedded as it was with childhood memories, and she very appropriately evoked the following lines from William Wordsworth's *Prelude*, a poem about trying to recover early memories of experiencing nature: "I watched the golden beams of light, flung from the setting sun, as they reposed, in silent beauty on the naked ridge."

But what, it might be asked, is a professor of literature doing meddling around with paintings and picture-taking? Isn't literature supposed to be concerned with reading and writing? Yes, but it's also true that others involved with exploring urban nature—biologists, historians, professors, activists—are heavily invested in the medium of writing. This valuably underscores the fact that for a broad reappraisal of nature in New York

to be realized, to really sponsor an increased land ethic in the city's constituents, the paradigm shift must be multidimensional, on all fronts—not just facilitated by the written word, but also by visual, even aural forms (every movement must have its beat, as was often said about the effective progressive actions taken in the 1960s). Finding beauty in the city streets and capturing it with a camera as the student did in the photograph offered here, and then linking the image to poetry that has been read and personally felt, strengthens a bond with a local place in a way that is different from pure descriptive writing or an analysis of local biodiversity.

Besides opening the practice of creative writing to include other media, I have found it correspondingly important to adjust the focus on what we read in a class as environmental literature. The young field of ecological literary criticism has been keen to explore its green roots and foundations, especially, as noted above, within the work of the Romantic poets, who valorized an emergent environmental perspective and were often deeply invested in science as a foundation for truth. (When Walt Whitman writes, "I sing the Body electric" in 1855, he really means it—his poetry is reflecting cutting-edge discoveries about the physiology of electricity in the human body.)

To think about the role of the imagination in these other areas of perception, apart from language—in sound, in sight—perhaps no

Romantic poet is worth more scrutiny than William Blake. Much of Blake's poetry is fantastically illustrated with his colored engravings, and reading it today is an experience akin to getting through an epic graphic novel. The ways in which certain images and colors and phrases serially recur in Blake has also been theorized as anticipating the hyperlinked logic of cyberspace. In Blake, the predominance of the image, often in tension with the word, becomes all the more relevant for our modern culture that is ever more visual and less textual.

When I teaching Blake's *Songs of Innocence* and *Songs of Experience* (1789–1794)—poems that were once sung by Blake, creating another sensory level of meaning to the texts—students immediately detect the recurrent themes of entrapment, loss, and mental enslavement that are juxtaposed against images of idyllic nature. Despite such pastoral tropes, many of the poems are peculiarly urban, oriented towards the fevered cosmopolitan space of London at the close of the eighteenth century, giving them relevance to city dwellers today. In the illustrations surrounding one of the opening poems of the series, tiny human figures are seen encased inside stiff foliage that looks a little bit like stained glass. In a later poem, "The School Boy," this border has become flexible, organic, and more green, and the figures leave the confined space of the design.

Symbolically, this transformation embodies (albeit inversely) the narrative of "The School Boy," which speaks much to anyone who teaches about the environment. I'd like to end by partially quoting the major stanzas of this poem, keeping in mind how Blake remained committed to living in a city that was being radically transformed by industrialization so intense and dizzying that many of Blake's Romantic colleagues (like Wordsworth and Samuel Taylor Coleridge) abandoned it for an idealized countryside. Blake writes:

> I love to rise in a summer morn
> When the birds sing on every tree
> The distant huntsman winds his horn
> And the skylark sings with me
> O! What sweet company
> But to go to school in a summer morn
> O! It drives all joy away
> Under a cruel age outworn
> The little ones spend the day
> In sighing and dismay.
> Ah, then at time I drooping sit
> And spend many an anxious hour
> Nor in my book can I take delight
> Nor sit in learning's bower
> Worn through with the dreary shower.

"The School Boy" reminds the teacher of environmental literature that always the best instruction lies away from the classroom, outside; it also reveals the power (for good and bad) that educational institutions can wield in shaping perception. Aldo Leopold saw higher education as one of the largest blocks in the way of realizing a collective land ethic; he felt that really "our educational and economic system is headed away from, rather than toward, an intense consciousness of land." Today, no huntsmen sound their horns in New York like they do in Blake's poem, and we don't have the same kind of lark that Blake was thinking of when he wrote in eighteenth-century London. But there are certainly thrushes, vireos, wrens, flycatchers, and brilliant scarlet tanagers—all species that

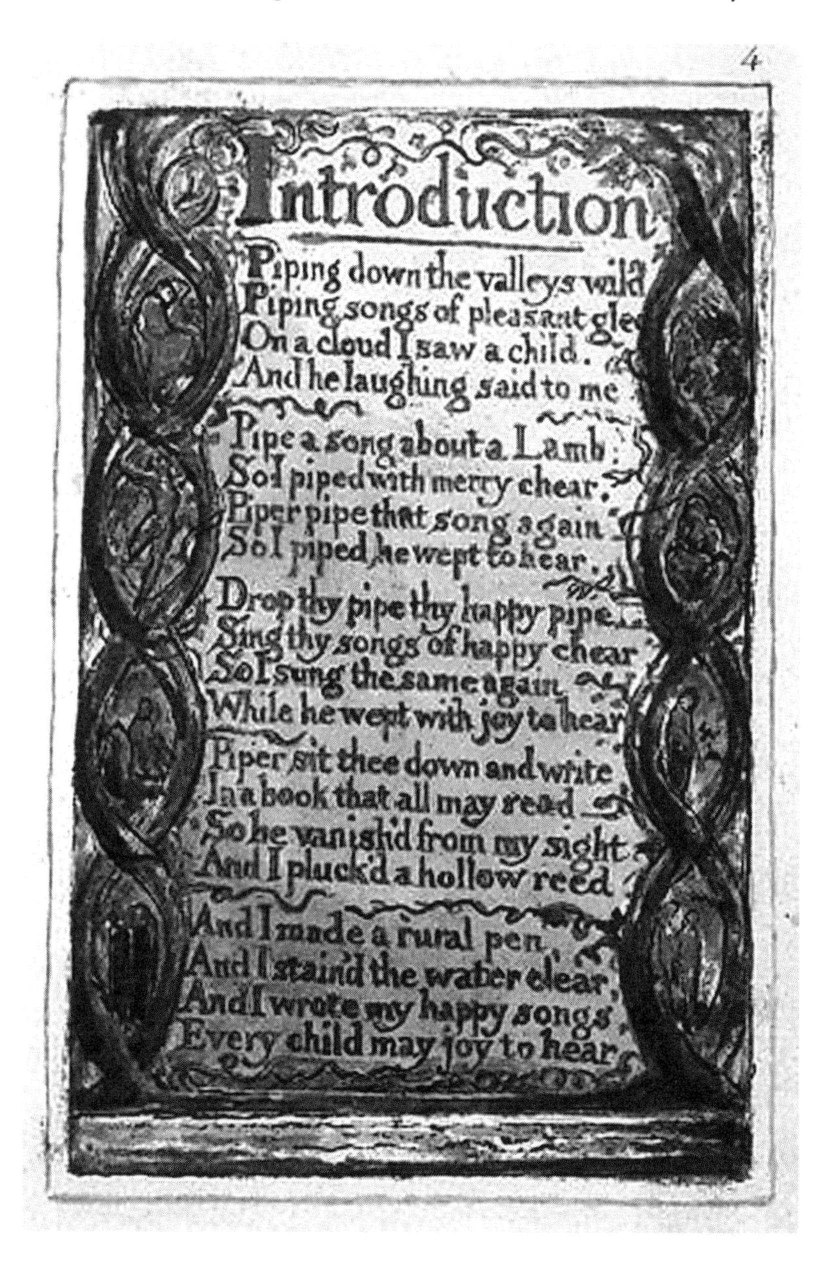

4

Introduction

Piping down the valleys wild
Piping songs of pleasant gle
On a cloud I saw a child.
And he laughing said to me.

Pipe a song about a Lamb;
So I piped with merry chear,
Piper pipe that song again—
So I piped, he wept to hear.

Drop thy pipe thy happy pipe
Sing thy songs of happy chear,
So I sung the same again
While he wept with joy to hear

Piper sit thee down and write
In a book that all may read—
So he vanish'd from my sight.
And I pluck'd a hollow reed.

And I made a rural pen,
And I staind the water clear,
And I wrote my happy songs
Every child may joy to hear

53

The School Boy

I love to rise in a summer morn,
When the birds sing on every tree;
The distant huntsman winds his horn,
And the sky-lark sings with me.
O! what sweet company.

But to go to school in a summer morn
O! it drives all joy away;
Under a cruel eye outworn,
The little ones spend the day,
In sighing and dismay.

Ah! then at times I drooping sit,
And spend many an anxious hour,
Nor in my book can I take delight,
Nor sit in learnings bower,
Worn thro' with the dreary shower.

How can the bird that is born for joy,
Sit in a cage and sing.
How can a child when fears annoy,
But droop his tender wing,
And forget his youthful spring.

O! father & mother, if buds are nip'd,
And blossoms blown away,
And if the tender plants are strip'd
Of their joy in the springing day,
By sorrow and cares dismay,

How shall the summer arise in joy,
Or the summer fruits appear,
Or how shall we gather what griefs destroy
Or bless the mellowing year,
When the blasts of winter appear.

persist and wait to be discovered as "sweet company," even as the winding of ubiquitous police sirens seems to punctuate our city air. Nurturing New York's nature requires that the imagination be set free to seek urban experience in the streets and to find moments of beauty there that are as profound as what we've read in books.

Nature in New York

A Brief Cultural History

Frederick Buell

The writers in this volume see themselves, for the most part, as coming to their subjects with both a special urgency and a sense of doing something almost outrageous, against the grain, counterintuitive. To that one can add a touch of that populist form of civic pride New Yorkers know and cultivate, the theatrical arrogance of doing it in the Big Apple, on the largest stage there is.

And there is truth in the contrarian outrageousness. For to write about nature in cities goes against the grain of at least four centuries of modernity, which has done its best conceptually and pragmatically to put nature and culture into opposite categories. It also goes against the grain of much local New York history—both its grim environmental history and its penchant for making fun of its "other," the heartland. Recognizing this, Anne Matthews investigates the return of wildlife to New York with the awareness that "of all U.S. cities, nature and culture here seem most spectacularly, insistently estranged." Robert Sullivan gleefully confronts these attitudes by investigating one of the most estranging parts of contemporary urban nature, its "dark side," the rat. Though conditions have changed as a result of 1970s environmental legislation, one of Devin Zuber's students puts the still-reigning wisdom trenchantly, with the sort of ironic hyperbole with which the City is most comfortable: "The problem with nature in New York is that there isn't any."

Discovering, recovering, representing, and disseminating awareness of the nature that actually still in fact does exist in New York is thus a challenge. It goes against both embedded habits of mind and practice that would oppose nature to culture and rurality or wilderness to the city, and it goes against deep-rooted local habits and attitudes of New Yorkers. But it also goes against many of the conventions of romantic and post-romantic nature writing, something that, until recently, would put attempts to write about nature in New York in one of the most marginalized corners of that genre. Nor can writers with this subject in mind find help in most traditions of urban writing; most urban authors (from the Epic of Gilgamesh to the present day) have with few exceptions accepted the divisions between nature and culture, wild lands and cities as imaginative and historical givens, just as vigorously as have nature writers. But today, as never before, the normativeness of these positions is under pressure to change, and writing about urban natures today stands to put one in the forefront of literary, intellectual, and also, one hopes, policy change. It is one version of the story of this dramatic change that I would like to tell in the following.

Be a John Muir or Henry David Thoreau in Manhattan—or in the Bronx, or Brooklyn, or Queens? That seems at first a quixotic ambition, to say the least. Even so versatile and empathetic a nature writer as Terry Tempest Williams writes about how, while working at the American Museum of Natural History in New York City, she went with a friend to see Pelham Bay Park in the Bronx, a spot that her friend cherished as an essential part of nature in the city. After traveling through ravaged urban neighborhoods, they arrived at the park on a "sinister" midwinter evening, to find a desolate beach, a vandalized pavilion, and a dead dog, a black Labrador, stiff on the beach beneath the setting sun, which itself resembled the tip of a burning cigarette seen through fog—and then had to wait for a cab, because trying to walk home would have been too dangerous. Herons and a flock of red-winged blackbirds eventually did rescue the day for her, but, still, the overwhelming impression was the melancholy of a place that had "little memory of wildness."

So, write about nature in New York? Wouldn't that simply mean having your most gifted colleagues (nature writers past and present) laughing at you (in contemporary literary journals, from across the river Styx)—or, worse, like Williams, indulgently pitying you? And wouldn't it confront urban writers with an almost irresistible temptation to say back something ironic about Williams's purist sentimentality and naïve shock at her short stay in the modern world? Wouldn't the writer who tried to get between these two adversaries be fatally marginalized by the assumptions of the very genre he or she had picked to write in? New

York chutzpah, of course, can do a lot to make space; but some odds even chutzpah cannot even.

But the situation is very different when one puts the enterprise of writing nature in New York in a contemporary context—or, more precisely, in three contemporary contexts, contexts that are in varying degrees environmental, historical, and literary-historical. Doing this not only does much to counter the marginalization of an urban nature writer, but actually even gives such a writer the edge—especially if that writer is aware of the fact that he or she is writing on such an edge.

The first context to point out is the most important and fundamental. In it, we encounter what is no less than the greatest of the world-historical transformations brought about in the post–World War II era. I do not mean the spread of new political systems or the modernization of societies around the globe or the transformations in cultures as well as infrastructures brought about by new technologies. I mean the growing awareness of the fact that the balance between human activity and natural process has changed in a way no previous society (save perhaps that of science fiction writers depicting the twenty-ninth century) could imagine. What had been unthinkable became, in a mere fifty years, routine knowledge that the majority of Americans came to consider simply one of the (usually more unpleasant) facts of life: namely, that human beings had the capacity to change nature everywhere and to do so quickly. The awesome size of charismatic natural formations—vast mountain ranges, impassable deserts—began, suddenly, to shrink, as did the still more awesome sense of the vast stretches of nonhuman time that surrounded and conditioned the paltry period of human habitation. What people did became as important, or more important. Formerly, people had been sensitively dependent on nature; now nature was also sensitively dependent on people. As human population and power increased, qualities attributed to nature, such as grandeur, size, and stability, disappeared into thin air. It was the atomic bomb, and the growing postwar prospect of nuclear holocaust, that introduced this notion: people could change everything everywhere at the push of a button. But this fear of nuclear devastation was quickly followed by extreme concern about a series of human environmental impacts, from runaway population growth, to the spread of toxic chemicals, to the destruction of ecosystems and the loss of species, to human-induced changes in the world's climate. Dipesh Chakrabarty has recently shown how quickly such a condition has become our new normal by suggesting that we now no longer inhabit the Holocene, but instead the Anthropocene.

Changes like these were shocking, and they were first conceived of in apocalyptic terms: the end of nature and thus society was at hand. Read Paul Ehrlich on the population explosion, Rachel Carson on toxic chemicals, or see the still-chilling-in-spots potboiler film *Soylent Green*, in which a pre-NRA Charlton Heston teamed up with Edward G. Robinson to investigate and then passionately mourn the fate of people and the death of plants and animals in a pollution-ravaged, climate-changed, totally urbanized world. Passionate idealization of nature was coupled with the sense that it was in extreme and imminent danger; and the combination of these emotions fueled the 1970s ecology movement—leading to the passage of major environmental legislation.

This grand, world-historical change in human-environmental relations, together with the extreme alarmism of postwar society's first cultural response to it, provides us with the first context for considering contemporary thought and writing about nature. The second context is what happened shortly thereafter. As convincingly awful as many of the prophetic visions of the 1960s and 1970s were, the end did not come in the decades predicted—nor has it come yet. The sun shines still, and birdsong is still audible. This failure of apocalypse to arrive is, of course, something that inspired and gave ammunition to the anti-environmentalist ideologues who, reacting against the environmentalist movement of the 1970s, began organizing themselves and working on many fronts to roll back that decade's tide of environmental concern. From the Heritage Foundation to the Wise Use movement, from the hectoring of Julian Simon to the counterscience of Bjorn Lomborg, the anti-environmentalist movement subjected environmentalists of a wide variety of stripes to withering fire. Suddenly environmentalists, not polluters, were the enemy, and a charming new lexicon of abuse emerged to characterize them: environmentalists were routinely called "chicken-littles," "doomsters," "toxic terrorists," and even "apocalypse-abusers."

But at the same time that this new political controversy about nature became prominent, something more interesting happened: a new intellectual controversy about nature and culture also arose, one in which many of the assumptions behind 1970s nature writing and philosophy were interrogated, criticized, and altered—altered for a time when human powers seemed to be more fundamental than nature's stability, and yet when the predicted apocalypse had not come. These changed assumptions about—or, in the language of the time, paradigms of—nature seemed, to many, to do no less than transform what the previous generation had seen as crisis into an opportunity for rethinking and retooling modernity in fundamental ways. The political and intellectual reactions to the failure of apocalypse to appear on schedule represent the second context I believe is important to keep in mind in thinking and writing about nature today. This context is less about a grand turn in environmental history than about the much more recent, local drama of ongoing cultural responses to environmental crisis.

The third important context is the most recent, and it is one still in formation. Despite the failure of apocalypse to arrive on schedule—and despite the screeds of the new conservative anti-environmentalist political opposition—and despite the emergence of new theorizations of nature and culture, ones that seemed to change crisis into intellectual opportunity, the great shock voiced by the apocalypticists of the 1960s and 1970s did not simply disappear into thin air. It did not go away because many of the grave threats they articulated did not simply go away; moreover, new ones became visible. Environmental crisis didn't simply disappear; in fact, it deepened. Concern about global warming (one of the planet's many stigmata depicted in *Soylent Green*) re-entered the mainstream; it has emerged from the era of anti-environmentalist debunking and disinformation as a major crisis and a decisive test of humanity's ability to act. Moreover, not just one but a variety of environmental problems have deepened, and new ones have been added. For example, the incidence of cancer in the developed world has shot up above the level Rachel Carson shockingly predicted, so that now over 40 percent of us will face a cancer diagnosis in our lifetimes. No person is left in the world, anywhere, who does not have traces of toxic chemicals in his or her body. A new specter of epidemic diseases has arisen along with globalization and global population growth. And anxiety about the dramatic increase in the rate of human-caused species extinctions has intensified, with scientists now projecting that our time will rival the great extinction events of evolutionary history.

An epigrammatic way of summarizing this third context is to say that, for us, now, apocalypse has not meant the end of the world, nor has it been successfully dismissed as a chicken-little deception, nor has it served as a spur to the conception of a liberating new paradigm; but rather, paradoxically, it has become a context into which we have lived our way more deeply than ever before. Apocalypse has become, as I argued in 2003, our way of life. Or, as Donella Meadows earlier put it, we now dwell beyond the limits. This sense of rising risk has been institutionalized in a wide variety of laws, regulations, scientific research, ongoing news stories, and human anxieties. In short, what once was apocalypse has become a way of life; what once was supposed to bring about the end of the world is now an important way in which the world proceeds on its uncertain path. We live under the shadow of mounting global problems and even more uncontrollably mounting perceptions of social-environmental risks—a situation that has become part of our daily life, even one of the pillars of it. Even though positively inclined reviewers may praise a new book in the still-continuing succession of environmental screeds by calling it "another *Silent Spring*," none will ever be that: Carson awoke an ignorant nation to terrible problems. We, by contrast, live in them consciously, talk about them, investigate them, argue over them as part of our daily fare.

That's depressing. But it's also quite positive, seen from another angle. A feeling of urgency does not have to mean passive despair—the feeling that we don't have the time or ability to do anything. Many things can and are being done, in smaller as well as larger arenas. And one of those arenas—one that, in the longer run, is fairly important—is changing old, apparently fundamental cultural conceptions about people and nature, culture and nature, and especially cities and nature. And this is precisely what the writers in this volume are, in the largest way, about.

Let me single out two specific points here. First, attitudes toward culture and nature are presently very much in flux. As nature has come to seem less and less the opposite of human society and culture, or even simply external to it, but instead everywhere (rapidly) shaped and reshaped by it, cities are gaining greatly in environmental significance and importance. One of the first to theorize this transformation was Bruno Latour, who argued in 1993 that we dwell not in nature, but in complex and varied sets of nature-cultures. Rather than being themselves unnatural aberrations, then, city environmental writers seem well positioned to be increasingly paradigmatic for nature's future, cities being places where nature and culture are most intensively and problematically fused. Second, working with these changed attitudes has opened new social and literary possibilities for activism on behalf of nature and writing about nature. If writers can help jolt or entice city culture out of its attempt to erase and separate itself from nature, more possibilities will be opened up for a world in which human culture and nature, cities and ecosystems, are increasingly impossible to separate. New York might help create future possibility as well as past disaster. If one can take the step beyond "there is no nature in New York" to "there are indeed many important forms of nature in New York," and also "there are many fascinating and instructive environmental as well as human histories to be recovered from the city's past," the world as a whole might be helped to work its way out of the straitjacket it is now in. It is no accident that contemporary environmental writers have seen the ecojustice movement—analysis and activism that responds to damage to nature and the consequent immiseration of marginalized people—as one of the most important ways in which environmental activism is being recreated in our postapocalyptic, natural-cultural era. Exploring the natures that have existed and continue to exist in New York can, in short, help people to think and act in relation to the fullest spectrum of entangled problems and challenges that they and their many kinds of nature, today, face.

So here is the story I would tell specifically about what lies behind the new literary encounter between nature and urban writers. I want to start with nature writing—and give an account of how it developed, thrived, and then began to mutate (often startlingly) in the postwar era. I'll take John Muir and Henry David Thoreau as my models for two dominant strains of thought in both nature writing and environmentalist action that became sharply foregrounded in the 1960s and 1970s activist eras. Then, wilderness (Muir) and wildness (Thoreau) became the trademarks of environmental concern. It is important to note that they did so even though the actual environmental activism of that time was in fact more various: it was as vigorously anti-corporate and anti-toxic it was pro-wilderness and pro-wild. But wilderness and the wild were the poster children of those decades: they were on all the calendars.

For nature writers, Muir established a fondness for the romantic-epic mode: for adventures out into wilderness, in which one would contact the original, the primeval, the pure, the unspoiled, and (above all) the Other: an Other that was often grand and sublime (wilderness as God's country, i.e., the country where God was) and always foundational. Experiences like these were also humanly transformative: one was awakened, changed, restored, enlarged.

Thoreau often struck many of the notes later struck by Muir, but his main interests were different. Writing in the romantic-pastoral rather than romantic-epic mode, Thoreau also left conventional society, but didn't go so far; and his most famous work, *Walden*, made dwelling in nature, not adventuring into nature, into the valued experience—an experience that also, over time, brought about a transformative change in the person undergoing it. Contact with the "wild"—with self-willed nature, vigorous and free—could occur on the edges of settlement just as well as out in the remote wilderness and could be, in fact, just as transformative.

Muir and Thoreau wove rich natural history writing into their dramatizations of themselves, and the results were prototypes for later nature writing. Human beings experienced semi-sacred encounters with the pristine or wild Other, encounters that were transformative for the persons involved, opening them up to full perception of the world around themselves and full experience of the nature inside themselves. To love wildness, for example, meant exploring, experiencing, and valuing it in the world; it also meant the liberating discovery of wildness in oneself. For Thoreau, this experience was not only a source of intense vigor, physically, and heroic nonconformity, socially; it was also a basis of spiritual freedom and literary genius. Shakespeare, for example, was for Thoreau not just a conventionally good writer but a timelessly great writer, because he (no mere follower of literary convention, but an originator) embodied wildness in his literary enterprise.

These notions of nature rode high during the early postwar period of environmental awakening and activism, a period that in 1970 produced the first Earth Day and led to a decade of crucial and substantial environmental legislation. The sacred Other was decisively, apocalyptically threatened, and protecting it seemed the most urgent, self-evident imperative. The result was a new flowering of nature idealism that showed itself in political passion and in a new wealth of nature writing and visual art, work that focused in particular on wilderness and wildness, on what seemed the most inviolable essence of nature. There was also an analogous conceptualization of nature in philosophy: ecocentrism and, more specifically, Deep Ecology became important philosophical-ethical movements. The literary theme of leaving society and the philosophical theme of rejection of anthropocentrism led to a recognition of the otherness of nature and a decentering of previously human ways of seeing things in favor of the awareness of a larger whole. And the result, in both cases, was awakening to one's fullest human potential as a part of that larger whole.

What I want to maintain now was that this widespread and intense embrace of the romantic nature tradition in literature and these developments in philosophy were quite appropriate to their moment. In the face of the discourse of environmental apocalypse that also flowered during this time, they represented both a potentially crucial brake to apply to a society run amok and a passionate formulation of the alternative, still seemingly within reach. And that this alternative did not seem merely utopian, but did seem actually within reach, is underscored by one of the most startling of the rich trove of factoids Anne Matthews includes in her book *Wild Nights: Nature Returns to the City*. She notes that 80 percent of the built world of the United States was constructed from the 1950s to the present. The world's population has also risen strikingly during that time—virtually tripling since the start of World War II. It was during the postwar period that the chemical industry blossomed and became an agent of global toxification; and technology took untold leaps during the same period of time. Capitalist reorganization of private and social life was in a primitive phase in the decades just after the war, compared to the present. In the early postwar years, the United States was still, to many, "nature's nation," and some sort of rollback to nature from the brink of apocalypse seemed a plausible ideal.

From both the urgency of coming apocalypse and the plausibility of a possible return to nature, then, arose the great emphasis placed on the wilderness tradition's values of transformative contact with the purity, pristineness, health, originality, primevalness, and, above all, Otherness of nature. Similarly, the Thoreauvian emphasis on wildness—which cast nature in the role of Other to society, depicting wild nature as self-willed, free, the opposite of what people experienced in society—came to seem the only pastoral literary act in town, even though an older literary tradition of the rural-agricultural pastoral was once dominant. This earlier pastoral tradition did not emphasize contact with the wild, but rather the wise human cultivation of nature with the goal of attaining a stable, fruitful harmony with it. Its ideal was the garden, a nature that humans were part of, not the sacred Other of wildness or wilderness. But this tradition was thrust out of the limelight by the urgencies of the 1960s and '70s.

With the 1980s, not only did apocalypse fail to occur, but the economy and technology began a great growth spurt. The result was that 1960s and '70s environmental politics started running into walls, and thought about nature in the United States began to take an important turn. This is a subject far too large and complicated for thorough discussion here; let me just cite one small but symbolic example to indicate the magnitude of the change. It is an especially ironic example: already in the 1970s, in one specific area, nature enthusiasm was exacerbating, rather than solving, the crisis of nature. Environmentalism's remedy was becoming the new problem. By 1973, Roderick Nash had added a new epilogue to his important study *Wilderness and the American Mind* (first published in 1967), warning that the popularity of experiencing wilderness had risen so much that the national parks created to protect it were overwhelmed by visitors, and what they were designed to protect was threatened, paradoxically, by its preservation. Other cultural critics—especially those who were part of the theory revolution that transformed and preoccupied literary study into the end of the century—began to spread the postmodern wisdom that these parks museumized nature, and that the wilderness they tried to preserve was not the primeval wild but a social construction. A consensus was growing that *all* areas of human life—from people's personal attitudes to the mediascapes they inhabited to the ideas they formed and the built worlds in which they dwelt—were socially constructed; and these attitudes were quickly extended to nature, especially since ecologists and environmental historians had become so clear in demonstrating the determinative effects on nature of human beings, from ancient to modern times.

One important result of this change in thought was that a number of qualities that 1960s and '70s activist ecologists had felt nature self-evidently possessed seemed to be suddenly undercut—and in a startling and provocatively contrarian manner. Most important, nature's once apparent Otherness seemed suddenly to be no more. In fact, it seemed to vanish in a number of different ways. Human beings had so encroached on nature, the apocalypticists had shown, that nature was no longer independent of people. Supplementing this sense of radical contemporary change were environmental historians' insights into how radically the earth's ecosystems had been reshaped over the course of human history, from the emergence of mankind, to the invention of agriculture, to the rise of modern society. Ideas of a timeless or pure nature were thus discredited. Finally, on the level of intellectual history, nature's "otherness" was undermined in an equally decisive manner when cultural theorists began to argue that this supposed attribute was not only a socially constructed ideology, but an ideology of surprisingly recent vintage. The "otherness" of nature was an ideology that had been put in place during the romantic period—and put in place so decisively that it seemed to *be* the nature of nature.

The most contrarian of theorists promoting this analysis then pointed an accusatory finger at 1960s and 1970s environmentalists. Their suggestion was that the socially constructed faith in nature's Otherness had led to ill, as well as good. For when special wilderness areas of the United States were designated in the early twentieth century, and when, in more recent times, nature parks were created in the Third World, the native occupants of these areas were removed from them, and the land was, in effect, *made* "pristine," or socially (re)constructed as pristine. Just as bad, these critics held (with logic, but not evidence), the idealization of these few preserved lands as "pure" and "untouched" might perversely license the more complete despoliation of everything else that wasn't "pristine." Accordingly, the ideals of nature's Otherness, pristineness, primevalness, and even the concepts of limits, balance, and health—ideals important to environmental thought, activism, and even regulation—were questioned and found by contrarians to be not only intellectually suspect, but also, when translated into politics and policies, potentially oppressive rather than liberating.

At best, these drastic revisions in thought about nature were part of a shift in balance from the ecology- and biology-based environmentalism that lay behind 1960s and 1970s fascination with wilderness toward a stronger and more visible emphasis on urban concerns, from human health to waste management and environmental justice. Less happily, however, this revisionist thinking sometimes meant an attempt by proponents of the latter to erase the concerns of the former—a new phase in old, often bitter debates between these groups, pitting Deep Ecologists against social ecologists and nature-based activists against social-justice advocates. For a while, environmentalism without nature seemed to a growing number to be a progressive possibility, not a paradox.

And at worst, of course, these ideas were very compatible with the rhetoric of ideologues opposed to environmental protections of all stripes, be they protections of nature or of people—ideologues who were simultaneously developing their versions of such ideas in well-funded conservative think-tanks and neopopulist activism. They were used, often quite flamboyantly, to roll back the legislative gains of the previous era. If pristineness was a myth, wilderness protection was delegitimized; if the supposed healthiness of nature was likewise a myth (diseases, after all, are a part of nature, and there are natural as well as manmade carcinogens), a chief standard for legislation against toxic chemicals and pollutions was undercut; if nature was never the original or primeval or Other it had been supposed to be (for, clearly, it had always been changing, and had changed in sync with human activity from the early migrations of hunter-gatherer peoples around the globe to present-day global modernization), then continued (even radical) reshapings of nature were not inherently suspect but simply business as usual. A new strip mine was simply evolution at work. These new half-truths then supported movements to roll back environmental protections in a wide variety of areas, including protection of wilderness, endangered species, and human health. Even friendlier to anti-environmentalist positions was the argument that the environmental movement was oppressive, not liberating.

Quickly, environmentalists of all stripes were vilified not only as "chicken-littles," but also as "eco-fascists," and "toxic terrorists"—people who were out to do bad things to the rest of us.

As I suggested above, this simultaneous development of cultural theory in the academy and right-wing, anti-environmentalist ideology in local and national politics did not just occur in a vacuum, but in sync with other kinds of change. Many other developments combined with it to make the nature idealism of the 1970s seem fatally wrong-minded, outdated or, at best, merely utopian. During the 1980s, human power to change nature was raised by a significant number of notches—and this very power, seen as apocalyptic in the 1970s, was aggressively restyled as positive, transformative, offering a prospect of untold possibilities. Globalization not only helped bring about the end of communism, but promised (to enthusiasts) a dramatically transformed world, one that would knit the whole species together for new economic growth and a new era of social synergy, not conflict. The economic boom in non-Western as well as Western countries—and occuring in the Pacific Rim even before in the United States—promised, to enthusiasts, a new era of growth and prosperity. Even in hitherto "exotic lands" tourism and development had arrived, and the notions of nature's otherness and of pristine, unaffected nature anywhere in the world seemed clearly passé.

At the same time, urbanism, rather than ruralism, became even more aggressively the norm for society as a whole. The time was rapidly approaching (indeed, it has just been reached) when the majority of the world's population would live in cities, and the megacity was becoming a regular feature of the developing as well as developed world. In the First World, cities themselves were expanding into conurbations, enlarged systems stretching far out from the traditional city itself. Cityness penetrated even far-removed rural areas—notably through the expansion of media that spread almost exclusively urban values in their broadcasts (*Lassie* doesn't exist even in reruns any more; instead, *Survivor* has so richly redefined nature in fundamentally urban terms, it's a subject for another essay on its own). As Burton Pike has argued, cultural visions of rurality and the pastoral had been created and propagated by city-dwellers for millennia; today, however, the urban media that had carried them has less place for them as alternatives or counterweights to urban lifestyles than ever before. The expansion of cities and cityness, like the global integration of the modern economic system, has thus helped give the quietus to old notions of pristine, "other" nature.

Perhaps most salient is the new technological boom, the subject of a thousand excited panegyrics. On the one hand, society was transforming old machinery into new megamachinery in the extractive industries, transforming farming from a rural activity into an industrial business, and transforming communications from a developed-world and nation-centered industry to a global network reaching even into Mongolia and the Amazon. People's ability to affect nature everywhere, and to do so quickly, seemed to jump to a new level. Quite literal examples of this were provided by such new technologies as bioengineering and nanotechnology. On the one hand, these new industries industrialized nature at the micro level, even as factory farming had industrialized it at the visible level. At the same time, technologies like these seemed to storm nature's last, shaky bastion of otherness: its lock on evolution. Truly, nature seemed to be no more.

These changes made some technophiles and postmodernists in the 1980s and '90s celebrate the fact that nature was now "over"—and to advocate nature's increasingly thorough and fundamental change through human intervention, not its protection from humans. New popular culture and even high literary-theoretical fetishes included the cyborg, the human-machine combination described by computer and robotics enthusiasts, and the chimera, the combination of species that visionary genetic engineers sometimes also celebrated. Such futurists as Alvin and Heidi Toffler and Julian Simon claimed that to be a pessimist was to commit a sin—and Simon, at least, showed no reticence in labeling environmentalists as the most sinful of the pessimists.

Not just cultural theory and anti-environmental ideology, then, but a new, runaway period of economic and technological development made the point clearly: there no longer was anything "outside" human determination.

But just because there may be no "outside," that doesn't mean nature is over; in fact, it increasingly means that nature is everywhere—everywhere inside—and that its condition is more immediately important to humans than before. For if it was remarkable that the threat of apocalypse seemed so quickly to be dispelled, it is also remarkable to realize how short-lived also were the enthusiasms that led to dismissing the claims of nature. For, with the end of the millennium, a large number of balloons seemed suddenly to burst, pricked by a series of difficult events. It is amazing how the failure of the dot-com bubble took the wind out of what seemed then to be our inevitable and wonderful techno-future. What's happened to all that brio today? It is amazing how quickly naïve celebrations of globalism foundered on the rocks of 9/11 terror and how quickly we seemed to be plunged into an unimaginably worse re-run of the Cold War. Francis Fukuyama had declared that history had ended; quickly, it seemed to start up again. Also chilling of new paradigms and enthusiasms was the rise to power of a Far Right–conservative alliance, some elements of which seemed to be attempting to turn the clock back to the era before the Scopes Monkey trial. More specifically, after all the transformation of crisis into opportunity by new social forms, new technologies, and new "paradigms" for thought, the nation elected what was, in environmental terms, an old-paradigm president, one devoted to the oil and extractive industries and disinformation and delay about growing environmental problems, not someone who celebrated the end of environmental issues because nature was at last "over."

So the 1990s disappeared as quickly as the Clinton-era budget surplus, and the new millennium became old, problematic, serious. The old fears of the end of the oil economy, utterly discounted during the go-go '80s and '90s, suddenly reappeared with a vengeance, as prices shot up at the pump and analysts proclaimed that peak oil production was at hand. Suddenly there was again a receptiveness in the wider public to thinking seriously about problems and crises—if only because we seemed to be plunged into a darker world of international terrorism, as well as utterly bogged down in the war in Iraq and other supposed attempts to deal with that threat. But also important was that people were finally beginning to listen to the scientific community, which had solidified its views about global warming as a major new crisis for our day. In fact, a number of environmental crises caused by rising human impacts are as megasized as this one, but the promotion of one crisis into genuine visibility was all that was necessary to help deepen the changed mood about what was, just a few years ago, touted as the magical, liberatory time of the twenty-first century, the New Millennium. And with this promotion, the environmental apocalyptics that had seemingly been debunked in the 1980s began returning in a new form: not necessarily as the end of the world, but as a sense of risk and anxiety that was being woven deeply into the fabric of daily life.

What then happened to 1970s nature writing during all this time—the time from the failure of apocalypse to the rediscovery of crisis in a new form? Were all its notions of nature irrevocably debunked and outdated? They weren't; nature writing, if anything, greatly expanded its niche in the publishing market. And it was surrounded and buoyed by the new

Photo by Ardythe Ashley

cultural movements of eco-literature, eco-criticism, and eco-theory, and the appearances of eco-literature courses in college curricula and eco-critics and theorists on college payrolls. Within nature writing, however, significant changes had begun to take place. Even as old attitudes and styles continued, the genre as a whole was beginning to undergo a powerful and significant sea change. Especially with the appearance of a new group of writers and a second generation of ecocritics and ecotheorists, nature writing has been responding to the pressures, theoretical and practical, on older conceptions of nature. It too has begun to use the tools of theory, including theory's exposé of concepts of nature as socially constructed, to craft new kinds of ideas about, literature of, and advocacy for the environment.

So now to my main point. Influenced by new conceptualizations of nature, by the changed balance of the relationships between humans and nature, and by the massive alteration of nature that has become visible as that balance has shifted, contemporary nature writers have begun to do important work in a number of new areas. In particular, they have begun to focus on exploring nature not as a force apart from human culture, but as intertwined with it. In this self-conscious expansion of the literature of nature, nature writing is turning into environmental writing and is merging with a variety of literary forms outside the nature tradition. In the process, its range is expanding. At the same time, the new public readiness to realize that society is living within a crisis brought about by human alteration of the planet's environmental systems has made many of these changes in nature writing feel especially urgent, important, and cutting edge.

In the new environmental writing, it's hard to deny that nature and culture are everywhere interwoven, not separable, both in the present world around us and in fully aware investigations of our history. No piece of earth in a polluted and developed world is pristine, original, unspoiled, unchanged, or wholly Other. To modify Pogo's famous statement yet again, we have met the environment and it is us. One common contemporary successor to Muir's wilderness writing, these days, is thus created by authors such as Colin Woodard and Mark Hertsgaard, who adventurously explore the world, summoning up natural history not to encounter the pristine, unspoiled, and Other, but on a quest to chart the extremes of environmental damage done to the world we occupy. A similar change has occurred in environmental autobiography; along with stories of liberating contact with nature's otherness come, today, stories of growing up with damaged environments, bodies, and even psyches.

But other contemporary nature writing emphasizes the genre's more traditional components of surprise and wonder, rather than degradation and claustrophobic anxiety. In a humanized natural world, gardening and cultivation as models for human-nature interactions, for example, have made a comeback. Such writers as Michael Pollan have focused on gardening, cultivation, and agriculture as today's appropriate metaphor for human-nature writing, and they track how long histories of human-nature interactions have helped create today's landscapes and biota, as well as social structures and attitudes.

More striking still is the emergence of writers devoted to discovering and fostering the presence that nature still has in the heart of human culture—in our cities, specifically. These writers have revitalized some of the oldest conventions of nature writing, but in important new ways. Writers in this vein appear in this book; additionally you can find their work in such anthologies as *City Wild* and in such individual works as Lisa Couturier's *The Hope of Snakes* and John Waldman's *Heartbeats in the Muck*. No longer simply screeds about what culture—and cities—have done to nature, urban nature writing is about the degree to which one can (re)discover nature in culture, both in the past and present. With those writers who are the most firmly based in the nature tradition, such as Couturier, these discoveries are accompanied by the fierce awakening of an ethics of care for wounded creatures and ecologies; a new intimacy comes from their connections with damaged natures. No longer, then, should city nature writers feel their oddity or marginalization, or wince at Terry Tempest Williams's pity; for the coming centuries, they may well be the most adventurous explorers and appropriate models to follow, not marginalized deviations. Now that the number of people living in cities has, for the first time in human history, exceeded the number of people living outside them, city nature writers are poised to come into their own.

Having told at length the story of nature thought and writing during the last forty or fifty years, I'll be much briefer about recent modifications in urban perspectives. Most important is a new wave of thought about modernization, called *ecological modernization*, that entered mainstream discourse in the United States with the impact of the United Nations' environmental summit in Rio in 1992, Al Gore's 1992 book *Earth in the Balance*, and Gore's candidacy for vice president in the Clinton administration. Old-style modernization, which focused solely on human social progress, had helped bring on environmental crisis by not considering its effects on nature at all. These effects were "externalized"—that is, regarded as outside modernization's bourne. But now that the effects of human growth and progress on nature had become so visible and had begun to constrain human life in so many ways, modernization's resources of rationality and practical, instrumental thought had to be focused on solving environmental problems, as part of future human progress and transformation.

Thus the new or ecological modernization reconceptualized environmental issues not as concern with an endangered, formerly pristine, natural Other, but as issues that cropped up everywhere, throughout the interior of industrial society: problems, for example, with a society that had formerly externalized environmental factors, rather than internalizing them; that had focused only on finance capital, not natural capital; that had pursued the development and exploitation of nature as if nature were inexhaustible and free, rather than finite; and that had valued technological advancement without regard to its effects on nature. True or ecological modernization, in contrast, required factoring environmental impacts into every human decision, not simply protecting nature from development or exploitation.

From this changed approach to nature a new environmental ideal has been born, one that fully acknowledges past and present human alteration of nature but seeks to make the resulting nature-culture mix functional, not dysfunctional. The new ideal was called *sustainable development* (and subsequently, in reaction to the fear that the word *sustainable* in the phrase would tend to be forgotten, it has come to be called simply *sustainability*), and, in its short life-span, it has helped give rise to research, thought, and policymaking in a wide variety of areas, from green industrialization and green building to green technology, green cities, and green social justice. Sustainability places primary focus on expertise and inventiveness in pursuing new technologies and brokering new arrangements between environmental experts, governmental policymakers, and industrial producers, rather than on ecosystems: nature needs to be successfully incorporated as a crucial part of the changing system of human-environmental interactions, not cordoned off as a sacred place apart from them. For cities, this has meant the emergence of a diverse group of new theorists, experts, researchers, policymakers, and practitioners, committed to paying a new kind of attention to human relations with the natures that still exist inside cities—attention that will allow resourceful people to make cities into sustainable fusions of the natural and the human, the ecological and the urban.

So, while nature as "outside the system" has disappeared, it has reemerged literally everywhere within it, from local development practices ("smart growth") to global security, from food production to automobile manufacture, from social justice to eco-justice, from urban architectural design to waste management. In this way, global warming is indeed an emblematic crisis for our time: clear from the start is that its disastrous possibilities impinge equally on the economy, on human health, on international tensions, on the impoverishment and displacement of peoples, on industrial techniques and technologies as on ecosystems, the health of the oceans and forests, the future of biodiversity, and the stability of the weather. It is a nature-society crisis, a crisis of an interwoven nature-culture. Further, global warming is no longer so clearly an apocalypse to come, but a slow crisis already in process—a condition of rising risk in which we are now embedded and anxiously dwelling. The "Eaarth" of Bill McKibben's recent *Eaarth: Making a Life on a Tough New Planet* is the irreversibly damaged planet on which we now actually live. And everywhere on this new planet, social processes and ecosystem problems are transparently, tightly interconnected.

As part of this (re-)emergence of natures in question "within the system," writers about cities have begun to pay closer attention to the ecologies and ecological histories inside their boundaries, even as nature writers have also embarked on what Robert Sullivan wittily calls a "reverse commute" to the city to do the same. The sort of attention that is paid by each group—by city writers newly concerned with nature, and by nature writers now commuting to the city—differs. Nature writers in the city focus, roughly speaking, on the ecological elements of cities, ones discovered through observation, adventurous encounter, and imaginative looks beneath the city's attempts to erase or repress its natures within, and they are marked by care for other species, for the experience of wonder, and by appreciation of the power of ethics and beauty. Equally, these writers seek to expand the human psyche to include (or to recover its innate connection with) nonhuman nature, as well as to gather knowledge about it. Urban writers, in contrast, focus on how environmental systems and other species have been woven tightly into human histories, and how their suppression and mistreatment also represents a distortion of social and human welfare, sanity and justice. And such writers are motivated by the goal of enhancing human life and robust urban community, rather than focusing primarily or exclusively on ecological health. The former writers, generally speaking, are far more ecologically literate and committed to the nonhuman world than are the latter. The latter, by contrast, are far more aware of social and cultural forces, how they work and how they might be reshaped, and far more committed to enlarging human welfare by taking fuller account of nature. In the process, the latter group typically eschews the pursuit of wonder and beauty that so inspires the former, in favor of drawing on the resources of irony and ironic sophistication, city dwellers' tough ability to live adventurously within a context that is excitingly, and even horrifically, out of balance.

In actual practice, an encounter like this between two traditions and perspectives means hybridization. In a time when city nature is an emerging subject—when writing about it is something new—authors will explore and audiences will feel the enterprise as different kinds of attempts at the hybridization of former opposites: of finding nature in cities, places where it supposedly wasn't. Thus some of the enlivening energy of chutzpah in the essays in this volume. Coming from the nature tradition, Anne Matthews, fully aware of the unlikeliness of her subject, brings to her exploration of wildlife in New York and her speculation about its future a sense of surprise, wonder, and even epiphany, qualities central to the tradition of nature writing. She urges her readers to "crawl ashore though great tangles of poison ivy, past rusted Chevys" at Jamaica Bay and "then hold up a truck mirror to observe the secret rookeries—but they're there, and flourishing. I had no idea." In the truck mirror, the nature epiphany ("I had no idea") of old is recovered. Similarly rooted in traditions of nature writing and representation, Devin Zuber sends his students out to find instances of the perspectives utilized in Caspar David Friedrich's nature paintings—and to find them in the contemporary city. He thus concocts a marvelous pedagogical task, one that imports a nineteenth-century trope of nature representation into the city, and the result is a hybrid, not an oxymoron.

Also primarily indebted to the nature tradition are Christopher Meier and Tony Hiss, who advocate making New Yorkers aware of the much larger and older address that underlies their street addresses—their "place of honor within the larger H2O region." Doing this will create "a greater sense of connection to the creatures and landscapes—the nature—that surround us" and will reveal "astounding beauty." In a more localist vein, William Kornblum focuses on a similar goal, seeking to open his students' eyes to the natural topography and environmental systems that lie concealed beneath the built world they inhabit. Bridging the gap between nature concern and urbanist commitment, Kornblum also seeks to make students aware of how dealing with, maintaining, and restoring these systems affects urban taxes and budgets, as well as the ecosystems themselves. In the process, he seeks to refashion his students' sense of community to include ecological citizenship and an urban land ethic as essential parts of it.

Betsy McCully's essay stands as another remarkable fusion of nature writing with urban humanism—a fusion that takes place as narrative and meditation, rather than as analysis or polemic. McCully, echoing Thoreau, goes to ocean-fronting New York (instead of lake-abutting Walden) to dwell consciously. In the process, also like Thoreau, she digs down where she stands and back into time to the substrate of tectonic nature in the city, even as she sharpens her senses to become aware of the wildness (from monarchs to horseshoe crabs) that lives in present-day New York. At the same time, she honors the histories of the city's building and rebuilding, its constant refashioning of its environments, and its present human identity as a city of immigrants, like herself, seeking to dwell, seeking to make a home. The migrating butterflies and migrant humans end up wonderfully complementing each other.

Much more robustly on the urbanist side of the spectrum are Robert Sullivan and Philip Lopate. Each in his own way sets out to deconstruct prior ruralist, wilderness-based ideas about nature. Sullivan satirizes in his essay the yuppie wilderness types he met in Oregon and argues that today "there is nothing that is pure," and to think that there is is to take "man out of the equation of nature, which, as I am attempting to show, is, I believe, misguided." Extending his critique of the nature tradition, Sullivan notes that he chose for his most recent book a mammal that, unlike the whale, "would garner no protesters, driving days from California." He chose the rat, and he comically invokes Thoreau to relate how he "went to the city to live with men, a lot of men, as well as women, in particular my wife, and, as I said, with rats." In an alley he selected there, he sought to explore and study, day after day, not the seasonal moods of Walden pond or the inspiriting value of the wild, but the largely unseen world of rats—and he does this as a revisionist successor to Thoreau, "no offense to him."

If Sullivan satirizes Thoreau, he also channels him: given Thoreau's great gifts as an urbane satirist, Sullivan, even as he distinguishes himself from Thoreau, rightly invokes him as tutelary spirit for a book ultimately about urban men (and women), as seen through the lens of the nonhuman species most symbiotic with them, yet one that they abhor. He thus seeks to make people look into the mirror of the nature they have created and see that it is, fascinatingly, repellently, them. In doing this, Sullivan supplements Thoreau with his own version of New-Yorkish wit, with its love of the grotesque, its gusto at the "yuck" factor of rats, and its comical invocation of "the dark side" of things.

Philip Lopate's self-conscious urbanism plays on older, more high-cultural cosmopolitan traditions than does Sullivan's. Lopate is both an informal essayist in the tradition of Montaigne and an urban flaneur, observant lounger about the city, full of insider knowledge of urban scenes and settings. Further, to this older urbanity Lopate brings the more contemporary urban perspective of Jane Jacobs, a perspective that values and empowers local knowledge and enterprise. Lopate, like Sullivan, also focuses on deconstructing past nature traditions, arguing that "environmentalist attitudes" that favor purist preservationism are utterly out of keeping with the instinctive ways of urban people and the lessons of urbanist thinkers. Against central planning devoted to preserving pristine nature's open space, Lopate poses Jane Jacobs's espousal of the "organic, higgledy piggledy" growth of New York City, and he argues specifically against parks made like museums, rather than for people, for the social and environmental value of infill and density, and for the valuation and preservation of the nonnatural in the natural—like a promontory in the East River that was the residue of a concrete batching plant. It had come to function like a natural formation, Lopate argues, but was nonetheless targeted by the EPA for removal because it was "unnatural." In suggesting this example, Lopate echoes William Kornblum's embrace of the legacy of Herbert Johnson, who used a bulldozer to create new nesting spaces for migratory birds—and thus created what became a wildlife refuge. But, unlike Kornblum, Lopate's focus is more on the preservation of the human than of birds.

David Rosane's essay, on the other hand, aims at being both more provocative and more distinctively hybrid in its form than either Sullivan's or Lopate's. On the one hand, Rosane writes with the exaggeration, aggressiveness, and impertinent virtuosity of the nature writer, polemicist, and philosopher Edward Abbey; like Abbey, Rosane also fuses concrete narrative and philosophical speculation, and does so in a capriciously informal style. In writing this way, both Abbey and Rosane invoke the tradition of the romantic nature essay—the tradition of Emerson and Thoreau. Perceptions are presented as new discoveries, and larger ideas as insights emerging fresh from the author's reflections and the narrative's events. At the same time, Rosane's subject is the urban world that Abbey loathed, and his essay explores the notion that today not only are culture and nature inseparably fused with each other, but an explicitly hybrid vision is also theoretically and socially necessary. Eschewing urbane irony and espousing outback hyperbole, Rosane argues that ecocentrism and anthropocentrism need to be fundamentally joined at the hip, and the ideals of human rights and the rights of nature must be interconnected. This perspective allows Rosane insight into the remarkable hybrid vision of New York nature possessed by the city's recent immigrants: people who—in Central Park, on the Bronx River, and on beaches in Queens—experience in a host of fresh ways the actually surviving natural histories of the city, even as they bring to the city both worldviews and languages in which nature and culture have not suffered the radical divorce from each other enacted by modern Western tradition.

Like Betsy McCully, Kelly McMasters writes about dwelling consciously in the city. But the city she represents is not one filled with wildness, but rather one in which damage has been done to both nature and people. McMasters writes about a neighborhood she (short on funds) moved into, focusing particularly on the vacant lot across the street from her apartment. She learns its history: it was first the site of the Brooklyn Union Gas Company, then an abandoned lot, refuge for the homeless and playground for children; then, its burden of toxic pollutants recognized, a fenced-off brownfield. A brief renaissance ensued, as the city tried to include the lot in its brownfields reclamation and development program—choosing, predictably, not to make it into an urban garden, but a lucrative housing development. But this plan failed when, to the consternation of neighboring property holders and even more the would-be developers, the lot was named a Superfund site. McMasters writes a gripping account of her personal experience in this plucky but shabby urban space, doing full justice to the local history, the colorful characters who populate the area, and the convoluted yet too-recognizable urban and environmental politics of the place. Though the end of the story represents both human and environmental failure, the essay refuses to escape into nature writing's wonder or urban writing's irony or black humor. Its tone is precisely crafted: it refuses to submit to the apparently intractable dilemma it puts so vividly and resolutely before us with its last sentence.

Dara Ross's essay "Corner Garden" is about the temporary fulfillment of what McMaster's narrative seeks: a richly successful reinhabitation and renaturalization of a vacant urban lot. Neighborhood residents—a local community of African Americans, immigrants from places like Belize and Nevis, U.S. migrants from Florida and the Carolinas—successfully reshape an abandoned lot into an urban garden. Its cultivation brings out otherwise untapped knowledge and expertise in a wide variety of people, who show themselves not only skilled gardeners, but also lovers of flowers, vegetables, and soil. In the space of the garden, the most intractably urban becomes also the most lushly natural: indeed, the garden is formed when impulses that kept fire escapes overflowing with marigolds, poppies, petunias, pansies, and impatiens overflow into and transform its open space. But as important, the garden then transforms the community that creates it. Time seems to slow; people come closer to each other; outdoor evening screenings of movies on a flea-market projector commence; an old moonshiner raises strawberries for wine; a wedding is celebrated. As in Rosane's essay, the under-recognized dilemma of migrant New Yorkers who dwell imaginatively in two or more environments at the same time becomes, in Ross's piece, the springboard of possibility, as old ways and old (environmental) knowledge flow into new (urban) space. Yet the essay does this in tones that come from a very different kind of nature writing than what Rosane draws on—a tradition as old as ancient Greece. Though one of Ross's subjects is ecojustice, her essay shows that the old, richly elaborated pastoral tradition, originally a sweet indulgence of the comfortable elite, is still very much alive (and alive in a form both vastly more real and also vastly sweeter than ever before) in the marginal spaces of New York City. Literarily, in her essay, the urban and the pastoral richly fuse—though this victory does not last in fact, as conflict between city and nature, and between the rich and the poor, reemerges: bulldozers appear to level the space, ironically for affordable housing.

Hybridization. Crossbreeding. Crosspollination. The work in this volume is a start, a real start. But it goes only so far. David Rosane and Betsy McCully probably achieve the most complete fusion of naturism and urbanism in this volume—drawing from both traditions for polemic and narrative, respectively. The writers here who are more simply indebted to the nature tradition (Matthews, Zuber, Meier, Hiss, McCully, and, to a degree, Kornblum) pioneer new ground, but hybridize less. They come to the city, mostly, to discover, uncover, recover the nonhuman creatures, ecosystems, and landscapes overlooked but still embedded here, despite centuries of separating nature from culture, nonhuman from human life, country from city. They come to the city with refreshing optimism and mostly politeness, and find what they find expansively, with wonder, beauty, and surprise. But, when all is said and done, in the process they find only a few people, and mostly nature.

The volume's urbanists also find themselves seeking something new, something newly hybrid. Sullivan and Lopate both reach toward the radical novelty of urban nature, even as they focus mainly on clearing space for it by critiquing past versions of nature. In his introduction to this volume, John Waldman does a fine, witty, anecdote-rich version of nature in New York as a source of outrageous comedy, black humor, and irony. In doing this, he presents the unusual perspective of someone who grew up in a thoroughly hybrid environment and therefore sees the trope of urban nature through a lens of nostalgia. In his hands, the literarily new becomes a condition that has long existed, but just wasn't culturally recognized until now. Kelly McMasters, however, goes further still, focusing on a nature that is so decisively urban that what to others is a startling hybrid has become simply the way things are. She has no need to look back on or dismiss older traditions of nature. Her environments are, and have been for years, a complex interweaving of biotic life, human alterations, vivid community stories, industrial and institutional history, and urban politics. And Dara Ross's urban garden extends this fusion of the natural and the human to its fullest extent: the pastoral becomes the urban, and vice versa, in what seems an expression of a new environmental and urban ideal.

Urban natures; nature-cultures; nature in the city; the city in nature; possible synergies; dysfunctional yet indissoluble relationships: nature-urban hybrids come in all these different forms. The essays in this volume clearly are innovative literarily; and that innovation in turn becomes a basis for the new attitudes, ethics, and politics we so urgently need today. Given what twenty-first-century society is up against, it is time for different people to squeeze their differently shaped shoulders together more tightly than ever before, in order to apply them to the same (albeit complex) wheel. This volume betrays some of the tensions involved in that process, but it makes a brave start.

Contributors

Frederick Buell is Professor of English at Queens College and the author of five books. A poet and cultural critic, he has written extensively about globalization and culture and about the cultural impacts of deepening environmental crisis. The latter is the subject of his *From Apocalypse to Way of Life: Environmental Crisis in the American Century*. He is now at work on a book about culture and fossil-fuel energy history, entitled *The Cultures of Coal; The Aesthetics of Oil*.

Tony Hiss, an author, lecturer, and expert on restoring America's cities and landscapes, became a staff writer at *The New Yorker* in 1963, and since 1994 he has been a Visiting Scholar at New York University, first at the Taub Urban Research Center and now at the Robert F. Wagner Graduate School of Public Service. In 2002 he also became a Fellow of the CUNY Institute for Urban Systems (CIUS). He is the author of thirteen books, most recently *In Motion: The Experience of Travel*, which will be published in paperback in 2012, and *H2O: Highlands to Ocean* (with Christopher Meier). Among his other books are the award-winning *The Experience of Place* and *A Region at Risk: The Third Regional Plan for the New York–New Jersey–Connecticut Metropolitan Area* (with Robert D. Yaro). His work has appeared in the *New York Times*, *Newsweek*, *Gourmet*, *The Atlantic*, and *Travel & Leisure*. He serves on the board of the National Parks of New York Harbor Conservancy. The National Recreation and Park Association awarded Hiss its National Literary Award for a lifetime of "spellbinding and poignant" writing about "how our environments, modes of travel, and other aspects of the American landscape affect our lives." The American Institute of Architects, New York chapter presented him with its George S. Lewis Award for three-and-a-half decades of writing that has made New York a better city in which to live.

William Kornblum conducts research in the areas of community studies, urban ecology, and environmental sociology at the Graduate Center, City University of New York. He has worked for many years on studies of parks, open spaces, and environmental issues in the New York metropolitan region and other urban regions of the United States and Europe. His research has figured in the restoration of Central Park and Times Square and the development of Gateway, Golden Gate, and other National Recreation areas. His volume *At Sea in the City*, about the waterways of New York, draws extensively on this research and environmental activism. He is active in a number of Graduate Centers, chairs the Board of Directors of a Manhattan homeless center, and is a member of the editorial board of *Dissent* magazine. Kornblum was the recipient of the 2005 Presidential Career Award for the Practice of Sociology from the American Sociological Association.

Phillip Lopate, a Brooklyn native, has written three personal essay collections—*Bachelorhood, Against Joie de Vivre*, and *Portrait of My Body*; two novels, *Confessions of Summer* and *The Rug Merchant*; two poetry collections; a collection of his movie criticism, *Totally Tenderly Tragically*; and an urbanist meditation, *Waterfront: A Journey Around Manhattan*. Lopate also has edited several anthologies, including *The Art of the Personal Essay* and *Writing New York*. His essays, fiction, poetry, film, and architectural criticism have appeared in *The Best American Short Stories, The Best American Essays* (1987), several Pushcart Prize annuals, *The Paris Review, Harper's, Vogue, Esquire, Threepenny Review*, the *New York Times, Preservation, Cite, Metropolis,* and many other periodicals and anthologies. Lopate has also taught creative writing and literature at Fordham University, Cooper Union, the University of Houston, and New York University. He is a Professor at Columbia University, where he directs the nonfiction concentration in the graduate writing program.

Anne Matthews is the author of a trilogy of place-studies, portraying distinctive American environments dealing with peril and change. In addition to *Wild Nights: Nature Returns to the City*, she wrote *Where the Buffalo Roam*, which was a Pulitzer Prize finalist in nonfiction, and *Bright*

College Years, which was named a *New York Times* Notable Book. She has taught at Princeton, Rutgers, Columbia, and New York universities, and her work has appeared in the *New York Times*, the *Washington Post*, *Outside*, *The American Scholar*, *Orion*, and the "Best American Science and Nature Writing" series. She now writes fiction and film with Will Howarth under the joint pen name of Dana Hand; their first collaboration, *Deep Creek*, was a *Washington Post* Best Novel of 2010.

Betsy McCully is a writer of nonfiction and fiction. Her book *City at the Water's Edge: A Natural History of New York* grew out of twenty years of nature exploration in her adopted city. She has given many talks on New York's natural and environmental history and has created a website dedicated to documenting the nature of the New York region (www.NewYorkNature.net). She is an Associate Professor of English at Kingsborough Community College of the City University of New York.

Kelly McMasters is the author of *Welcome to Shirley: A Memoir from an Atomic Town*. The book was listed as one of Oprah Winfrey's top five summer memoirs and is the basis for the documentary film *The Atomic States of America*, a 2012 Sundance selection. Her essays, reviews, and articles have appeared in the *New York Times*, the *Washington Post Magazine*, *River Teeth: A Journal of Narrative Nonfiction*, *Newsday*, *Time Out New York*, and MrBellersNeighborhood.com, among others. She is the recipient of a Pushcart nomination and an Orion Book Award nomination and teaches at mediabistro.com and in the undergraduate writing program and Journalism Graduate School at Columbia University.

Christopher Meier is the co-author (with Tony Hiss) of *H2O: Highlands to Ocean*, an assessment of fourteen indicators on the environmental health of the New York/New Jersey metropolitan region, the H2O region. He has continued the work of this book by participating in discussions and forums in and around the H2O region. While he works in the software industry, he remains a "region-builder"—a champion of natural areas—by avocation.

David Rosane is a freelance naturalist and educator who has worked primarily as a science writer and journalist in France and as a Cornell University research associate teaching tropical ornithology and ethnobiology to minority undergraduates in the Venezuelan and Dominican rainforests. He recently lived in New York City as a naturalist-in-residence employed by the late Ted Kheel and his Nurture New York's Nature Foundation, working together with CUNY faculty to teach urban ecology to continuing education and College Now students. He currently spends time between Paris and his home in Vermont writing educational media about the environment and learning organic farming. He created and ran the HIDI outreach foundation, a small humanitarian program that worked with Ye'kuana natives of southern Venezuela; this career-long project has since been terminated by the Hugo Chavez government.

Dara Ross teaches English to recently arrived immigrant students at the Brooklyn International High School. She was the first African-American Peace Corps volunteer to serve in Mongolia, where she worked as an English teacher in a tiny village. Her writing has been published in *Essence* magazine and on Peacecorps.gov. She loves to teach, write, and make crafts.

Robert Sullivan is the author of *The Meadowlands*, *A Whale Hunt*, *Rats*, and, most recently, *The Thoreau You Don't Know*. He has written for magazines, such as *The New York Times Magazine*, *The New Yorker*, *New York* magazine, and *Vogue*. His latest book, *My American Revolution*, is a consideration of the Revolutionary War and the landscape of New York City.

John Waldman joined the faculty of Queens College as Professor of Biology in 2004. For the previous twenty years he was employed by the Hudson River Foundation for Science and Environmental Research, most recently as Senior Scientist. He received his Ph.D. in 1986 from the Joint Program in Evolutionary Biology between the American Museum of Natural History and the City University of New York, and an M.S. in Marine and Environmental Sciences from Long Island University. Among his research interests are urban aquatic environments, historical ecology, and the ecology and evolution of fishes. Dr. Waldman has written more than eighty scientific articles; an award-winning book on the environment of New York Harbor, *Heartbeats in the Muck: The History, Sea Life, and Environment of New York Harbor, Revised Edition* (Fordham, 2013); and another on marine phenomena and coastal pleasures, *The Dance of the Flying Gurnards*. He also has written and edited a number of other books and is an occasional contributor to the *New York Times*, *Yale Environment 360*, and various periodicals.

Devin Zuber is an Assistant Professor for American Studies, Literature and Swedenborgian Studies at the Graduate Theological Union (GTU) in Berkeley, California, where he serves as a faculty member for the Ph.D. Program in Art and Religion. Dr. Zuber received his Ph.D. from the City University of New York, where in 2009–10 he received the alumni and faculty award for most distinguished dissertation. Before coming to Berkeley, Devin was the in-residence Eccles Fellow for American Studies at the British Library in London and taught for three years as an assistant professor of American Studies at the University of Osnabrück in northern Germany. His scholarship has appeared in *American Quarterly, Religion and the Arts,* and *Variations.* He is presently completing a book on American environmental aesthetics, as well as a chapbook that features an interview with the performance artist Marina Abramović. He has held fellowships at the Glencairn Museum of World Religions and the Smithsonian's Hirshhorn Museum of Modern and Contemporary Art, and he serves on the board of directors for CARE, the Center for Arts, Religion, and Education, at the GTU.